Silver Dart Canadian Aerospace Studies
Volume III

Weapons in Space
Strategic and Policy Implications

Wilson Wong

Centre for Defence and Security Studies
The University of Manitoba

The Centre for Defence and Security Studies
University College
The University of Manitoba
Winnipeg, Manitoba R3T 2M8
Canada
www.umanitoba.ca/centres/defence

Silver Dart Canadian Aerospace Studies

Published by the Centre for Defence and Security Studies
in Winnipeg, Manitoba, Canada.

Weapons in Space: Strategic and Policy Implications
Wilson Wong
Silver Dart Volume III

First published in May 2006

> On February 23 1909, the young Canadian engineer John Alexander Douglas McCurdy lifted the Aerial Experiment Association's *Silver Dart* off the frozen waters of Bras d'Or Lake on Cape Breton Island to register a historic first: the first powered flight in Canada, and in the British Empire. McCurdy gets the credit in brief historical accounts, but his famous flight in the *Silver Dart* was the culmination of a concerted collective effort under the inspired leadership of Alexander Graham Bell. In fact, his team of Americans (including the aviation pioneer Glenn Curtiss and Thomas Selfridge), and Canadians, F.W. 'Casey' Baldwin in addition to McCurdy, had already made several flights in the United States in the *Silver Dart* and three earlier flying machines. Utilizing the first water-cooled engine in an aeroplane and the adaptation of ailerons for more controlled, stabilized flight, the true significance of the *Silver Dart* rests in the area of aeronautical innovation and design.

© Centre for Defence and Security Studies, 2006
Managing Editor: Lasha Tchantouridze

All Rights Reserved

Printed by Contemporary Printing, Ltd.
in Winnipeg, Manitoba.

Legal Deposit
National Library of Canada

ISBN 0-9780868-0-5
ISSN 1712-9524

Silver Dart Canadian Aerospace Studies

Series Editor: Dr. James G. Fergusson, *the University of Manitoba*

Silver Dart **Canadian Aerospace Studies** offers theoretical perspectives, strategic analysis, technological assessments, and historical evaluations of air and space issues in the Canadian and international context. It is designed to provide a specialized outlet for scholarly research in the aerospace field, with a specific focus upon Canadian defence and security. Silver Dart seeks to bridge the gap between academic and policy analysis on issue of aerospace concern from an inter-disciplinary perspective.

Members of the Board of Advisors:
Dr. Carl A. Christie, *the University of Manitoba*
Dr. Allan D. English, *Queen's University*
Dr. Chris Pogue, *ISIS Solutions Canada*
Brigadier General (Ret.) Joe Sharpe
Dr. Elinor Sloan, *Carleton University*

Contents

Weapons in Space: Strategic and Policy Implications

Acknowledgment .... *vi*
List of Abbreviations and Acronyms .... *vii*

Introduction 1

Chapter One: Space Weaponization Concepts

    Space: Operational and Political Considerations 7
    Militarization and Weaponization 15

Chapter Two: Military Space Potentials

    Information and Sanctuary 35
    Space Denial, Survivability, and Space Control 38
    Hegemony, High Ground, and Common Ground 45
    Ballistic Missile Defence 49
    Resurgence of Nuclear Weapons in the Tactical Role 52

Conclusion 63

Appendix One: Some Enabling Technologies 67

Appendix Two: Revenge of the Lighter-than-Air Craft 79

References 87

About the Author *95*

Acknowledgment

The militarization and weaponization of space is for many a "push-button" issue, clouding this important issue in rhetoric and emotion. Visions of Armageddon or enslavement by foreign influences represent the disaster scenarios for opposing extremes in this and other security and defence debates. The debate over the use of weapons against space targets and placement of weapons in space is filled with hopes, fears and a fair amount of hyperbole. Maintaining a cold dispassionate viewpoint on the subject is for some a disturbing venture. Yet this is the path that this book seeks to take, and I've had some help in this endeavour.

First of all there is the faculty and staff of the University of Manitoba's Political Studies department, and my classmates to thank. Within this academic environment I've actually found a place to develop my interests and more importantly achieved some degree of success. These are also the people who share some responsibility for the number of trees consumed by me in getting this far.

Specifically I'd like to single out the members who participated in my Master of Arts defence in April of 2005. These are the people who deemed my original Master's research paper (with some additional polish) fit to have a proper cover and binding. Oddly enough, Dr. Fergusson had the most difficult of time getting me to actually stop writing the original paper. Thank-you for allowing me to keep fiddling with this collection of thoughts and allowing its transformation into *Weapons in Space*.

Finally there are my co-workers who had to put up with me and my writings. Tim, Nicholas and Daria who have spent hours going over earlier revisions looking for the grammatical errors I am known for. This debate is guilty of creating words and jargon that otherwise have no place in the English language. Many times it has been necessary to assure them that most of the things discussed here are either real or thought close to being real enough to warrant funding. For helping me fit all this together, I am grateful.

Wilson Wong
March 2006

List of Abbreviations and Acronyms

AAA – Anti-Aircraft Artillery
ABL – Airborne Laser
ABM – Anti-Ballistic Missile
AFDD – Air Force Doctrine Document (the US)
ASAT – Anti Satellite (normally referring to weapons)
BMD – Ballistic Missile Defence
C3I – Communications, Command, Control, and Intelligence
DEW – Directed Energy Weapons
FOBS – Fractional Orbit Bombardment System
 (normally referring to the specific Soviet strategic weapons systems)
GEO – Geostationary or GeoEarth Orbit
GPALS – Global Protection Against Limited Strikes
 (successor to Brilliant Pebble)
GPS – Global Positioning System
 (normally referring to the US Navstar system)
HANE – High Altitude Nuclear Explosion
HPM – High Power Microwave
ICBM – Intercontinental Ballistic Missile
IOC – Initial Operating Capability
JDAM – Joint Direct Attack Munitions
LEO – Low Earth Orbit
MAD – Mutually Assured Destruction
MEO – Medium Earth Orbit
MIRACL – Mid-Infra-Red Advanced Chemical Laser
NASA – National Aeronautics and Space Administration
NFIRE – Near Field Infrared Experiment
OST – Outer Space Treaty (1967)
PGM – Precision Guided Munitions
RF – Radio Frequency
RMA – Revolution in Military Affairs
 (normally refers to current US transformational concepts)
SAM – Surface to Air Missile
SDI – Strategic Defence Initiative
SDIO – Strategic Defence Initiative Office
SBL – Space-Based Laser
SMV – Space Maneuver Vehicle
SRMS – Shuttle Remote Manipulator System (also known as Canadarm)
UAV – Unmanned Aerial Vehicle
USAF – United States Air Force
WMD – Weapons of Mass Destruction

Introduction

The importance of space to Western security and prosperity has brought about a clear need to protect Western space assets. The best means to providing such protection is somewhat less clear. Safeguarding the satellites that provide critical services to consumers and warfighters on Earth leads to the issue, but not necessarily the reality, of "weaponizing" space. In the debate over the future military use of space, the technical and political distinctions between militarization and space weaponization can be best described only as being "fuzzy." Indeed, depending on ones point of view, current and near-term military space endeavours have already crossed the threshold for weaponizing space.

Space is undeniably a political realm. The political nature of space ranges from the battles to fund the smallest of space experiments to the global issues of how, and even whether, space is to be used to the "benefit of all peoples."[1] The importance of space to national security has seen space militarized from the very beginning of the Space Age. Technology and the capability it bestows form only part of the reality that will compel or dissuade the weaponization of space. Events in space, for the near term at least, are dependent on terrestrial politics. Despite the strong image of space being the realm of science and precision technology, space policy is in the end about policy, with all the consequences and compromises that policy entails.

As technology continues to develop, it permits more policy options for the realm of space, and it seems unlikely that political conflict will disappear, the continuing pace of technological development will only create more options for taking warfare into the highest of "high-grounds." It is the convergence of technology with political utility that will determine whether the stationing and usage of weapons in space becomes a reality. Such a convergence between political will and engineering ability, however, is less than totally assured.

Space weaponization is not a simple "yes or no" matter. There is a perception that the debate is between, "idealistic, liberal arms control enthusiasts who oppose all weapons against warmongering militarists who never saw a weapon they didn't like."[2] Instead, the topic of space weaponization encompasses many different roles, capabilities, and technologies, each with different impact upon a nation's security. Even the definition of what constitutes a space weapon is largely the result of the political art of labelling. Blanket statements concerning the deployment of space weapons largely ignore the complexity of the issue. Such all-encompassing dismissal or support of placing weapons in space obscures the real utility and consequences of specific systems. Over time, these nuances may very well make such all-or-nothing positions seem naïve or dysfunctional.

Admittedly, a nuanced view on space weaponization would tend to support the idea that weaponization will occur at some time. Even the most limited deployment strictly for "defensive" purposes would constitute for some a breach of the "taboo" against weapons in space. Sacrificing ideals such as a "weapons free" outer space may be necessary for providing security in future. Practical policy on space militarization must be flexible to accommodate the constraints of the security environment.

While not totally inconceivable, it is somewhat hard to imagine a near-term scenario where the first generally recognized acts of warfare in space do not involve the United States (US). Indeed, even if two minor powers managed to cause wilful destruction in and/or from space against one another, there would be major ramifications for US space policy, investments and activities. It is through focussing on the implications for the US that this discussion on space policy, as it relates to national power, takes place.

It is the agencies responsible for the national security of the US that have benefited the most from space technology. The military use of space is of consequence to those concerned with the preservation and enhancement the US's power and status in the world, and it is impossible to discuss the rationale for and against the weaponization of space without centering on the last of the Superpowers. Presently, it is the US that has the greatest lead in space utilization, and the greatest perceived reliance on space systems.[3] Also, while scientific discovery and technological development rarely stick to plan, the United States also has the greatest capacity for space research and development.

At the same time, the US is not the only nation active in space. Like sea power and air power, the US does not hold a comprehensive monopoly on space power along economic, military and scientific dimensions. Taking notice of US success, other nations are developing their own space capabilities. Despite the marvels of the space race, space power is only in its infancy. The proliferation of space capabilities provides not only to the US, but also to its allies, competitors and potential adversaries incentives to weaponize space. It is unlikely that the US (or indeed any other global actor) will have its military use of space unchallenged indefinitely. It would be similarly foolhardy to expect such challenges to remain isolated to the diplomatic arena.

Current discussion on space weaponization is also largely framed as a "what if" debate. The near-term technology to field such capabilities simply does not match the hopes and fears found in the rhetoric from all sides of the weaponization debate. A consequence of this gap between capability and rhetoric, alongside space's association with science-fiction, creates much of the "giggle factor"[4] associated with arguments about space policy.

Nonetheless, space technology is well embedded in the background of modern society. Space technology is so embedded into the background of modern life that it is often only noticed when something goes horribly wrong. Indeed warfare in space for some sides of the space weapons debate would be about as horrible as intentional events can be. Regardless, the growing recognition of just how critical space infrastructure is to the modern world brings to attention the need to protect it, and marks it as a potential vulnerability.

As the recent suborbital flights of *Scaled Composite's* private manned spacecraft have demonstrated, space access may no longer remain the realm of big government.[5] Key enabling technologies for the mass weaponization of space, such as low cost routine space access, are also on the wish list of purely commercial space ventures. While the prospect of space tourism does not immediately translate into a specific national security threat, it should be remembered that many consumer level technologies have dual-use potential. There is nothing inconceivable about a hostile powers using civilian equipment against one another. Perhaps it is not a matter of whether technology will allow the placement and operation of weapons in space, but a matter of when the technology will arrive. It is likely the prerequisites for full weaponization will exist before it becomes routine to place holiday goers in orbit.

Arguably, the technology to cause property damage in space already exits today. Existing space systems, such as ground to space high energy laser experiments, have latent "attack" abilities.[6] Though at times laughed at, the fielding of a dedicated space combat ability is simply an engineering project. That is not to say that the technology is common or even exists in a practical form, only that the basic principles needed to develop space based weapons systems are more or less well understood. Given a large enough budget and time, anything within the laws of physics can be made a reality. It is just a matter of putting all the various engineering solutions together in a package that will actually do the job. This is true for all space-faring nations though, once again, it is the US that seems to be closest to deployment of a practical space warfare capability.

At the same time, just because something can be done does not mean it should or will be done. There are situations where hardware achieves initial operating capability (IOC) after the requirement for its development loses relevance, or is found to have never existed in the first place. The trend towards lengthened research and development times for contemporary military equipment does not help matters. History abounds with technological "dead ends." Reasons other than reality have allowed, and continue to allow, such projects to be funded well after dysfunction was recognized.

There is also the problem of technology simply being oversold. Being physically possible and being engineered to be possible are two different things entirely. In trying to avoid the act of deploying new but obsolete equipment, there is certainly justification for pushing the technological envelope. However if technological solutions cannot be found in a reasonable amount of time, if at all, then such efforts go to waste. Though valuable experience can be gathered from projects like these, building something before the technology is truly available will only result in unfulfilled expectations.

Neither spaceflight nor weapons procurement are innocent of wasting funds on projects with faint hopes of success, or projects that yield nothing of tangible value. Critics of the combination of spaceflight and armaments argue that the technology is not cost effective for the capabilities being described, and that these capabilities being promoted are not suited for the missions being (or expected to be) undertaken. Another variation on this theme is the political fallout from the technologically possible. Deploying a space weapons capability after a great expense could result in a loss of security for the US and a loss of stability for the world in general. As such, a technological success may well be dysfunctional for its actual purpose.

Politics, the decider of what, where, and how resources are to be allotted will ultimately decide on the need for and timing of space weaponization. The claim that an objective "expert" viewpoint will best decide many important policy matters such as the environment or security is rarely put to the test. In a democracy, perceptions on need, cost, opportunities, and capability have just as much, if not more so, to do with policy outcomes than objective technocratic thought. Perceptions are, of course, clouded by emotions. As the growing space weaponization debate has shown, there is quite a bit of emotion involved in what some see as one of the last environments "untainted" by man's darker tendencies.

Just like nuclear technology before, future developments in space technology will have ramifications on international politics. The technology needed for space weapons likely will (or already has) reached the point where, like nuclear weapons, it cannot be "undiscovered." Therefore, once the technology exists, it becomes a matter of dealing with the problem of what to do with it. It is a question of what developments in the military use of space will confer the most security. That is not a static calculus as the state of the world and state of technology are constantly changing. Even the variables to consider are subject to debate. The potential that space weapons bring is expressed in more than just technical terms. Aside from the harsh realities of physics, the perceptions on need for space weaponization will be the most important deciding factor.

Notes

1. Department of State, "Treaty on Principles Governing the Activities of States in the Exploration and Use of Outer Space, Including the Moon and Other Celestial Bodies," 27 January 1967.
   <http://www.state.gov/t/ac/trt/5181.htm> (2004).

2. Karl P Mueller, "Totem and Taboo: Depolarizing the Space Weaponization Debate," (Paper based on presentation given to Weaponization of Space Project of the Eliot School of International Affairs Space Policy Institute and Security Policy Studies Program, George Washington University, 3 December 2001),
   <http://www.gwu.edu/~spi/spaceforum/TotemandTabooGWUpaperRevised%5B1%5D.pdf> (2004).

3. Benjamin S. Lambeth, *Mastering the Ultimate High Ground: Next Steps in the Military Uses of Space*. RAND.
   <http://www.rand.org/publications/MR/MR1649/MR1649.ch5.pdf> (2004).

4. Lt. Colonel, USAF, Martin E. B. France, "Planetary Defense: Eliminating the Giggle Factor," *Air & Space Power Chronicles*, August 2000.
   <http://www.airpower.maxwell.af.mil/airchronicles/cc/france2.html> (2004).

5. British Broadcasting Corporation, "SpaceShipOne rockets to success," 4 October 2004. <http://news.bbc.co.uk/1/hi/sci/tech/3712998.stm> (2004).

6. Michael E. O'Hanlon, *Neither Star Wars Nor Sanctuary*, (Washington, DC: Brookings, Institution Press, 2004), 27.

CHAPTER ONE

Space Weaponization Concepts

Space: Operational and Political Considerations

Outer space, while being many things to many people, on a technical level, is more readily defined than the concept of "weapon." The Karman Line at an altitude of 100 kilometres[1] above the surface of the Earth is generally regarded as the boundary of space. Aside from the crews of space capsules and space shuttles, a handful of test-pilots, including as of the end of October 2004 two civilian pilots on privately funded flights,[2] have earned so-called astronaut wings by flying aircraft (specifically rocket planes) up to and beyond this altitude in suborbital flights.

The higher the altitude, the progressively thinner the atmosphere gets. At 100 kilometers, the notional boundary of space, there is hardly enough atmosphere to support flight via aerodynamically generated lift; however, at this boundary of space, the atmosphere is still too dense to allow for stable un-powered residence in space. At a high enough altitude to minimize aerodynamic considerations such as air resistance and with enough energy expended, a spacecraft can orbit. An object in orbit, or freefall, has in addition to falling,[3] a sufficient velocity component in a direction perpendicular to a line between it and the centre of the celestial body it is orbiting. This momentum perpendicular to the direction of "falling" causes the object to continuously "miss" the celestial body it is orbiting. The Law of Conservation of Momentum defines the orbital velocity for a spacecraft for a given altitude.[4] On the other side of the balance between momentum and gravity, a spacecraft that achieves momentum for escape velocity, will break orbit and be flung away from the body it is orbiting.

If the orbit is low enough, the atmosphere encountered will produce enough drag to have a pronounced effect. Even a faltering orbit involves tremendous speeds. The faster the object, the greater the air resistance or drag produced. At minimum orbital altitudes there are still enough atmospheric molecules to make drag a problem. Larger objects as a consequence of presenting more surface area generate more drag, and therefore have a shorter lifespan in orbit unless periodically boosted.

Air resistance also generates heat from friction. A satellite in a decaying orbit will likely break up due to extreme heat and other effects of air resistance before it loses most of its "forward" velocity. Relative to sea level, the atmospheric density at the boundary of space is almost imperceptibly thin. However, when combined with the speed needed for the lowest orbit, air resistance will have a noticeable effect. As the untimely re-entry of Skylab in

1979 demonstrated, the density of the atmosphere can vary unexpectedly to the detriment of objects in lower orbits.[5]

There is no sudden boundary between air and space. As the atmosphere gradually thins with altitude, several layers are defined. From the surface of the Earth these are the Troposphere, Stratosphere, Mesosphere, and Thermosphere layers.[6] The majority of human activity thus far is limited to within the Troposphere. In between the operating envelopes of even the highest flying aircrafts and the lowest orbiting satellites is the Mesosphere. Sustained operations by air or space craft do not occur in the Mesosphere,[7] which lies between the Thermosphere and the Stratosphere layers. Indeed, wing or aerodynamically supported flight is barely conducted in the upper portion of the Stratosphere. It is only at a sufficient altitude within the Thermosphere layer, that it is possible for a spacecraft to maintain enough velocity for at least a few orbits.[8]

Despite the creation of the term "aerospace" by the US Air Force (USAF),[9] negligible to non-existent atmospheric conditions found at orbital altitudes make outer space a totally different operating environment from the lower levels of the atmosphere where the majority of pilots are currently confined.[10] Though comparisons are made to air power, transplanting the dog-fighting component of air superiority into space is only a fantasy. Aerodynamic support and the lower speeds (kinetic energy/momentums) of aircraft flight share little in common with the careful management of tremendous energies involved with controlling a spacecraft in orbit. The laws of Newton and Kepler[11] mean that current spacecraft have less freedom of movement than mammoth warships at sea.

The difference in density between the medium of outer space and the realm of "air forces" has further implications over claims that warfare in one is readily applicable to the other. Space systems have to be tailored for operations in that environment, just as submarine weapons are tailored for operating in the dense medium of water. For instance, the Airborne Laser (ABL) and Space-Based Laser (SBL), two projects with potential application as space weapons, cannot share the same laser technology. Such technologies "might sound similar, but the differences are so fundamental that the ABL laser device is not traceable, scalable, or leveragable to the SBL laser device in any meaningful sense."[12] Differences in operating environments have resulted in significant engineering divergences beyond the basic concept of attacking targets with high energy lasers mounted on free flying platforms.

Space is a hostile environment. Lack of atmosphere allows orbiting and the ready propagation of laser energy, but also creates severe operational considerations. Without the protection of Earth's thick atmosphere, equipment is exposed to intense heating by the sun, extreme cold when in the shadows, and the effects of transitioning between the two extremes (often simultaneously on the same spacecraft). Aside from visible light, spacecraft are directly exposed to other forms of radiation from the sun and other sources. Certain applications require that satellite orbits encounter or are totally within the Van Allen Radiation Belts. These satellites require extensive shielding, not found on most others. Producing equipment that can function under such demanding and harsh conditions (hardening) is clearly expensive. As such, space is damaging enough to equipment and budgets without having to prepare for wilful acts of violence.

In addition, there is no convenient way to perform maintenance on equipment in space. There are few examples of on-orbit repair and construction, and all that have so far occurred have been rightfully hailed as extraordinary examples of skill, planning, and funding.[13] The ABL is expected to be serviced regularly in the comfort of hangars on the ground. In light of existing launch technology, SBL will have to be built to function without regular (possibly any) maintenance throughout its operational lifespan. Furthermore, if deficiencies are discovered in an ABL aircraft after deployment, the system can be adjusted or sent back to the contractor with relative ease on the ground. The heroic effort needed to fix the defective Hubble Space Telescope after it was launched only serves as a warning for the need to get things right when it comes to expensive space platforms like the proposed SBL. This level of engineering perfection only adds to the already considerable problems of space operations.

Of all the technical difficulties of space, none are as great as the simple problem of access. Perhaps the greatest argument against near term stationing of weapons in space is that space launch capabilities are still far from routine, with the related consequence of also being uneconomical. Extreme cost is no problem if the threat is great enough. During the Cold War, the huge arsenals of nuclear weapons each superpower deployed were not optional, but critical to continued survival according to the deterrence doctrines that, for the most part, guided strategy. However the Cold War and the arms race it entailed are over. Funding for new defence capabilities now faces different criteria of need.

Every launch into orbit requires a great amount of energy. As mentioned before, it is not just a simple matter of going straight up.[14] The speed "forward" needed to keep a spacecraft from hitting the Earth in the continuous freefall of orbit is much greater than that needed to loft it on a suborbital (ballistic) flight. The velocity needed varies with altitude/radius of the orbit. However a minimum of about 25 times the speed of sound (Mach 25) is the usual benchmark for attaining Low Earth Orbit (LEO).[15] Higher orbits require much more energy. It should also be noted that the great energies needed for satellite placement must be released in a precise manner so that the satellite actually gets to its proper orbit.

Currently, it is only through a controlled explosion of a chemical reaction based rocket that space can be reached. While there has been much speculation over the ability of technology to scale back the army of ground crew and bureaucracy[16] (producing leaner launch operations and cost savings along much the same philosophy as found in New Public Management), the energies involved cannot be reduced. Advanced launch concepts such as using air breathing engines partway are only more efficient ways of generating such energy. Other proposals such as thermal rockets powered by beaming energy from ground based power plants or more controversially onboard nuclear reactors directly heating a working fluid highlight the energy requirements by seeking to tap energy sources with greater potential than onboard chemical reactants.

All major powers, and some medium ones, such as Brazil,[17] have potential indigenous space launch capability (as to whether a specific nation decides to invest in such capability, and do so successfully is another matter). However the cost of space launch remains a problem despite increased competition. Governments and large aerospace companies, with government subsidized research and equipment, have thus far been the dominant players in getting satellites to where they need to go. In the words of Greg Klerkx:

> This catch-22 – a small market with high prices because it's a small market with high prices – has been the great barrier for the entrepreneurial space community, and the reason why those brave few who have embarked upon private launch-vehicle development have invariably failed.[18]

Continued sticker shock has led to the end of many goals dreamed of in the early Space Age, such as manned exploration beyond Earth orbit. Even what some would call the nightmare of warfare in space must first submit to fiscal realities. If for no other reason, the near term placement of weapons in orbit is uncertain due to the great expense such endeavours would entail.

Space technology's current "experimental" status is a large factor in its expense. Launch vehicles and satellites are far from being routine items even for military procurement. Spaceflight at present has more in common with launching an expedition than the sortie of a warplane. It is often claimed that by going into space more often the incentive to develop lower cost space access (among other necessary cost saving technologies) will finally break the present "experimental" model of spaceflight. This "Industrial" model of space technology implies something less delicate than today's space technology. Robust and routine weaponized space capabilities may actually have to be a reality prior to any major expansions of the commercialization of space.

There have already been attempts to operate in space on "industrial" levels. Though financial failures,[19] the huge constellations of low orbiting communications satellites planned under Iridium, Globestar and similar commercial ventures of the 1990's and early 2000's do show that the concept of mass producing scores if not hundreds of spacecraft is not beyond reach. At the same time these commercial venture were active, the US Strategic

Defense Initiative Office (SDIO) was contemplating the production and launch of hundreds of kinetic energy missile defence interceptors for the Brilliant Pebble and later scaled down Global Protection against Limited Strikes (GPALS) systems. With the creation and maintenance of these satellite constellations, there was a perceived need for robust low cost space access. Within both military and commercial worlds, several schemes to lower the preparation time, personnel needed, and ultimately costs were proposed. A few of these launch vehicle concepts continue to be promoted now after the collapse of so-called "satellite boom" of the late 1990's.

Alternatives to space technology took away much of the demand for mass satellite constellations and attending launch capacity. Ground based cellular technology has reduced the commercial viability of satellite phones, fibre-optics have proven to be better links for most of the wired world than satellite broadband, and the Clinton and subsequent Bush Administrations have redirected much of missile defence development towards ground, sea and air based systems. These are not the first examples of space projects being superseded by more down-to-earth equivalents, nor will they likely be the last. Just like the stagnation in airship development, technological developments elsewhere may supersede the technology needed for routine orbital operations.

Enhanced space access would also be necessary for some passive military space doctrines. Having the ability to rapidly replace satellites would arguably reduce the vulnerability of space systems by simply making each orbiting satellite individually less important by virtue of having a replacement in waiting. Potentially this could mean the wholesale loss and replacement of dozens if not hundreds of satellites to restore capabilities that can be provided only from space. Even if a technological defence was thought possible, the costs associated must be weighed against that of being able to overwhelm an attacker with satellites.

Perhaps it is more accurate to say that only certain types of space warfare attractive to the US defence community face the serious problem of getting to orbit. Orbiting is not necessary to qualify as a space system or a space weapon; however, the proposal of an orbit-based weapon tends to draw much more attention to the subject of space weaponization than a ground based alternative would. The controversy over the orbit-based Near Field Infrared Experiment (NFIRE) program is in large part due to it taking place in space with equipment retaining its "kill vehicle"[20] label from its previous use in ground based missile defence research.[21] The placement of a stripped down kinetic kill vehicle into orbit has led to charges that NFIRE is a move towards the deployment of an orbiting space weapon, and hence a major breach of the taboo against such weapons.[22]

Related to the problems of getting into a stable orbit is the problem of manoeuvring once orbital velocities are achieved. A component of Newton's

First Law of Motion is that an object in a particular motion tends to stay in that particular motion. If it was not for the constant gravitational force between an orbiting satellite and predominantly the Earth, the satellite's motion would be a straight line tangential to the orbit. Application of an unbalanced force to a satellite would result in a change in its orbit: lowering, raising, or changing its orbital plane. Momentum changes may be applied slowly, result in gradual changes to the orbit; for instance, atmospheric resistance present at lower orbital altitudes slowly wears down a spacecraft's momentum, eventually causing it to fall from orbit. However for dramatic manoeuvres, such as to avoid an unsophisticated ramming or unguided projectile/mass debris type attack, large amounts of energy must be applied over short time spans.

Except for a few exotic propulsion schemes (solar/beamed energy type sails, electrodynamic tethers),[23] thrust must be produced by consuming onboard propellants (either chemical reactants or working "fluids" expelled by means of pressure, heating, or acceleration by electric and/or magnetic fields). Propellants are finite, and represent mass that must be carried up into space by the launch vehicle as part of the satellite. For these reasons, a satellite's ability to change its orbit is severely restricted. Counteracting natural forces that would gradually move a satellite from its useful planned orbit already places heavy demands on these limited "fuel" supplies. This does not leave many options for orbit changes for military reasons beyond keeping a satellite in its operational orbit. Operational decisions to make use of in-orbit refuelling to extend the life of expensive long duration satellites or to use cheaper frequently replaced satellites, both fall into the earlier problems of getting things into orbit.

Objects in stable orbit tend to stay in orbit. This includes the aftermath of an attack in space. In the space community, there is much concern over the growing amount of so-called "space garbage;" the remains of satellites that have failed, spent launch vehicle stages, and other artificial debris from the various space programs. Warfare in space, unless carefully designed to somehow remove the target from orbit (uncontrolled re-entry in the atmosphere or more fancifully capturing the target intact by vehicle capable of surviving re-entry), will leave hazardous debris for other spacecraft. Any totally disabled satellite will circle the Earth uncontrolled with the potential of impacting some other spacecraft with disastrous effect.

More destructive methods would result in a cloud of debris that will closely follow the original orbit. Smaller debris are difficult to track, and due to the smaller surface area presented to the fringes of the atmosphere, may last in orbit longer than an intact dead satellite. There may be an almost unimaginable amount of space just within the useful orbital paths, but increased occupation by uncontrolled objects in orbit increases the chance a spacecraft will encounter one of these hazards.

Subatomic particle impacts occur with much greater frequency than those of natural and artificial space debris. Impacts from subatomic particles, the dangers of radiation, are a particular problem in space. Subatomic particles penetrating electronic equipment among other damage, cause changes in electrical properties of materials critical to satellite operations.[24] Exposure to radiation is an ever present hazard to operating outside of Earth's atmosphere.

Radiation exposure is not uniform in space. Natural events, such as solar activity and even interstellar events, from time to time increase the amount of radiation to which a spacecraft is exposed. In addition, where a satellite's orbit takes it can dramatically raise the amount of radiation exposure. The Van Allen Radiation Belts are of particular note as radiation hazards.

The Earth's magnetic field forms two donut shaped regions of high charged particle density, commonly known as the Van Allen Radiation Belts. By redirecting and "trapping" charged particles, the Van Allen Belts provides for Earth a significant amount of protection from natural radiation sources in space. Particles enter and leave the Van Allen Belts naturally forming equilibrium. Normally, the two belts more or less maintain their position and shape relative to the Earth. In between the two belts, there is the "slot;" a region where satellites may function without much consideration for expensive shielding and hardening against radiation.[25] The low density of charged particles in the slot is a product of the Earth's magnetic field's tendency to move particles into the Van Allen Belts. Within the Belts however, the higher density of charged particles increase in the rate they strike satellites when compared to outside of the Belts.

Events such as high levels of solar activity, or nuclear detonations at high altitude (away from shielding provided by the atmosphere) can "pump" the amount of particles entering the Van Allen Belts to overtake the rates at which they leave. This "pumping," if severe enough, can increase the radiation (charged particles) that a satellite orbiting in the slot is exposed to. Given the right distance from the atmosphere, a large enough nuclear event, (an intentional nuclear explosion or particularly unfortunate natural set of circumstances) may, "'pump' the belts with an overflow of electrons, artificial belts can form in the slot that can disrupt satellite electronics."[26] These excess particles take time to dissipate naturally from the environment and pose a hazard to any unshielded spacecraft till levels return to those prior to the "pumping" event.

With all these dangers from just being in space itself, there is also the problem of recognizing an attack in space. Current limits on space monitoring give rise to fears that the US (or any other satellite operator), would be unable differentiate satellite failure caused by natural events and those by less than overt hostile acts. For both sides of the space weapons debate, inability to at least monitor objects in space increases the risk of an accident turning into a diplomatic matter. It should be mentioned that a known Anti Satellite (ASAT) capability would tend to raise suspicions that the unexpected failure of militarily important satellites was an intentional act of hostility. With the existence of "latent" ASAT capabilities, such fears are not set all that far in the future.

In a very real sense, space, like the oceans, is a "common" territory. The physical realities of orbiting spacecraft have led to the notable political differentiation with terrestrial mediums of space not having (generally recognized) borders. Orbiting satellites, spacecraft and potentially weapons platforms (fractional or otherwise) circle the Earth with no regard for the international boundaries below. As for the position taken by certain equatorial nations in the Bogotá Declaration of 1976, claims of sovereignty up to geostationary (also known as GeoEarth, GEO or Clarke[27]) orbital altitudes are ignored by physics and by the international community.[28] The laws of Newton and Kepler make it impractical to enforce the concept of sovereign airspace beyond the boundary of space.

Operationally, the space or slot along an orbital path occupied by a satellite does constitute "territory." A better analogy would be that spacecraft, like ships and aircraft in international territory, have the freedom to navigate with a safe separation between them. Satellites have owners, and in the case of nationally launched and otherwise owned spacecraft, have territorial rights like military and other nationally owned vessels. Infringing on the freedom of navigation for a ship is usually regarded as an unfriendly act, if not *casus belli*. The care taken to ensure that nothing intrudes into the space occupied by the International Space Station or other crewed spacecraft points out that there is something akin to "right of way," in outer space.[29]

In the 1990's the idea of a "right of way" in space was tested during several disputes between the Indonesian satellite operators of Palapa B1 and operators leasing a geostationary orbital slot allocated to Tonga.[30] Initially this was just an example of disputed ownership (and royalties Tonga demanded from Indonesia) of resources in the new frontier, which escalated in 1992 with the hazardous placement of a second satellite by one of Tonga's clients into the slot occupied by Indonesia.[31] Later in 1996, a second client of Tonga who by that time had sent its own satellite into the disputed space had its satellite "jammed," by the Indonesian satellite. While both incidents involving this particular Indonesian satellite are not considered "warfare," at the very least they do suggest that space is ripe for belligerent activities. Perhaps the most worrying aspect of this example of satellite interference is that the nations involved in disputing rights to the orbital slot are not great technological or space powers.[32] It also highlights the ease with which a satellite may be subject to interference.[33]

Developing internationally agreed upon procedures, rules and norms for safe practical space operation (and to avoid further international incidents in space), is an ongoing process. Analogous regimes exist for both air and sea mediums though there are limits as to how far nations, and especially global powers, are willing to go. Well meaning efforts to prohibit air warfare and to limit naval weapons have clearly failed as these two mediums are quite heavily weaponized.[34] Despite this, opponents of further space militarization make claims that space can avoid similar levels of militarization.

The desire for further international regimes governing space is balanced against concerns that such agreements would have the effect of reducing a nation's perceived security. The recent Commission to Assess United States National Security Space Management and Organization (more commonly known as the 2001 Space Commission) welcomed such international regimes, as long as they are not intended to reduce US security. In their report the Commission stated that

> To protect the country's interests, the U.S. must promote the peaceful use of space, monitor activities of regulatory bodies, and protect the rights of nations to defend their interests in and from space.[35]

In the face of the practical concerns of the last remaining superpower there are many other voices in the international space regime debate. As a common territory, there are certainly those outside the US who present arguments over their entitlements to the benefits of space despite their own lack of capacity to do anything in space. The Bogotá Declaration mentioned earlier goes beyond optimistic claims of space for the benefit of all, to an attempt to define sovereignty well into space. Orbital slots are at present not allocated by their relation to borders on Earth. At present, history has limited the ability for most equatorial nations to control and appropriate the common territories just offshore, let alone overhead.

Militarization and Weaponization

In any discussion on the space weaponization it is necessary to go over the distinction between "weaponization" and "militarization." For the purposes of this discussion a distinct difference between the two terms needs to be maintained. The "militarization" of space can, in one sense, be thought of as all the reasons militaries are in space, including (but not necessarily) actual combat. The addition of a combat capability, defined as the ability to apply coercive effect to another nation's property and personnel, acting in or from space is then the "weaponization" component of the military's use of space. As Michael O'Hanlon puts it,

> Although space is becoming increasingly militarized, it is not yet weaponized – at least as far as we know. That is no country deploys destructive weapons in space, for use against space or Earth targets, and no country possesses ground-based weapons designed explicitly to damage objects in space.[36]

Militarization short of "weaponization" has been going on for several decades now. The military presence in space is extensive ranging from satellites that perform surveillance, missile and nuclear event warning, to communication and navigation. Currently, short of errant spacecraft ramming other spacecraft or points on the surface of the Earth, there are no readily available means of applying coercive force in the traditional sense. Even without actually shooting and otherwise causing damage in the exercise of national policy, the abilities conferred by satellites in space are militarily valuable.

Another perspective on space militarization and weaponization is the differentiation between active and passive means of dealing with military threats in and from space. Passive means do not actually involve attempting to directly counter or interfere with a target or with an attack. In its strictest sense, passive means for protecting space assets take into account the existence of threats and are designed to avoid vulnerability. Making a satellite invisible to an opponent would be an example of a passive defence.

Active means of military space operations on the other hand require direct intervention on an opponent's equipment and interests in and from space. Force application with the intent of permanent disablement or outright destruction of a spacecraft certainly would fall into this category. This includes force being projected from within the atmosphere against orbiting spacecraft. Active military operations against space assets would clearly be an example of space warfare.

In between, there is certainly room for systems that may be considered one or the other depending on political needs. Temporarily inhibiting or degrading an opponent's capability is included in what are termed as offensive counterspace operations. The 'Five Ds,' "deception, disruption, denial, degradation, and destruction – describe the range of desired effects when targeting an adversary's space systems."[37] Depending on how critical the services being stopped are, no distinction may be made by the target of the first four of the "Five Ds" from the final "D" of destruction. Blocking of signals and/or a satellite's field of view, while usually described as being passive in nature, are steps that in a tense situation may be regarded as hostile acts that require a military reply.

It is important to note that this distinction between "militarized" and "weaponized" ("passive" and "active") is not shared by all. For some, the weaponization of space debate is part of a bigger debate over the use of coercive force in general. Moreover, there has, at the very least, been a blurring of the role played by some military space systems, making the distinction problematic. In some respects the integration of space systems with terrestrial weapons, Space Force Enhancement as it is termed by the US military,[38] pushes the boundaries of what may be considered peaceful activity in space.

Space and other modern systems have arguably reduced much of the fog and friction of war. Low cost precision weapons and other capabilities enable comparatively small forces to accomplish military tasks which once required much larger investments. Space systems provide much of this force multiplier effect for the US military and other like-minded forces. These advanced or transformational capabilities are described by many as a Revolution in Military Affairs (RMA). Max Boot goes as far as referring to this war fighting mindset as the "New American Way of War."[39] The chief characteristic of this new way of war is a break from the attrition warfare of the two World Wars. The argument for this type of precise warfare also implies a desire to reduce casualties for all sides. RMA capabilities promise the ability to apply exacting force only to the target, reducing both collateral damage and overall investment in the process. Space borne assets are responsible for much of the precision, communications, information superiority and other components necessary for such an economic use of force.

In the past, when satellites and the immediate recipients of their services were few, space technology was viewed as a strategic concern. Satellites would generate the maps needed for the location of targets, and later on, additional geological measurements to improve the accuracy of missile guidance systems. As late as the 1990's many conventionally armed cruise missiles, such as the earlier model BGM-109 Tomahawks, used complicated terrain matching and inertial guidance systems, which relied on satellite generated maps for their function.[40] Mapmaking is still second-hand-use, as these "smart" weapons were largely left to their own devices after release.

Today, military satellites have a direct military effect. Initially satellite services were utilized by individual weapon platforms. Munitions have recently gained direct access to these services as well. For instance, the current versions of the Tomahawk cruise missile now use Global Positioning System (GPS) satellites to assist in navigating to their targets.[31] Tactical munitions have benefited greatly from this direct access to satellite provided services.[42]

Aside from the already mentioned point of view denouncing all forms of coercive force, the fact that satellites now are critical to many tactical weapon systems makes the space weaponization debate seem superfluous in the eyes of a few analysts. It could be argued that, by having direct participation in the target engagement cycle by an asset in space, the whole "system" for target destruction forms a tactical space weapon. The combination of off-board sensors (whether they be on a satellite or an aerial vehicle) with weapons delivery platforms has been often labelled a "reconnaissance-strike complex."[43] Certainly, by being a critical data-link or guidance system on a weapon, these space systems would tend to go against the idealistic notion of "space for peaceful purposes."[44] These space systems allow for many of the capabilities key to the military transformation being promoted among

Western defence communities and those of many competitors.

In other words, the RMA underway today has, at the very least, resulted in orbiting satellites becoming critical elements of tactical weapons systems.[45] Space based systems not only allow individuals to know their position to within a few meters, but also to call down satellite guided fire by simply calling in the target's position.[46] In this sense, the current RMA allows soldiers to wield space weapons; weapons whose expected performance and safe operation (to the one wielding them) are dependent on space systems. Indeed with the proliferation of the "joint" concept for all things military, space becomes a critical part of a total "joint" battle space encompassing all assets involved in a conflict from individual infantry to satellite constellations providing communications and navigation services in real-time.

Conceptually, the Joint Direct Attack Munition (JDAM) and other weapons that rely on space systems for their effectiveness are not generally considered examples of space weaponization. Instead, these systems are usually categorized as Precision Guided Munitions (PGM's) or "smart" weapons. The lineage of guided munitions can be traced back to Second World War era weapons like the German Fritz-X remotely guided bomb,[47] well before spaceflight was a reality. Their present integration with space systems is largely overlooked: "policymakers and populations clearly do not consider the deployment of such weapons [JDAM] to constitute space weaponization."[48]

In the other direction, it is not necessary to "attack" orbiting satellites directly to achieve their neutralization. Elimination of critical ground infrastructure by infantry would put the continued operation of many space systems in jeopardy. Ground based interference of ground based satellite receivers, such as handheld GPS jammers meant to interfere with the nearby operation of GPS receivers was brought to attention as a potential US space vulnerability by among others, the 2001 Space Commission.[49]

In terms of the effect on the target, it is unimportant whether the weapon was launched from a ship, aircraft or orbiting platform. However the concern appears to be less about effects than the mechanism of destruction itself that is of importance to policy makers and the general public over the space weapons issue. Under most, but not all, analysis, the military use of space has not yet included the full deployment of systems that apply force in space or from space themselves. Again, it is useful here to point out that the actual kill mechanism of contemporary space supported weaponry, the actual missile or bomb, is not stationed, deployed from, or destroys targets in space. According to John M. Logsdon,

> The use of signals from Global Positioning System (GPS) satellites to guide precision weapons to their targets is akin to the role played by a rifle's gunsight. But there are not yet space equivalents of bullets to actually destroy or damage a target.[50]

The whole matter of traversing space between terrestrial launch point and target is another controversial area in the space weaponization debate. There is some disagreement over where weapons that temporarily exit the atmosphere (such as ballistic missiles) fit into the whole militarization of space. Even the primitive German V-2 missile spent a portion of its short ballistic flight path in space. Consequently, by some definitions, space is already a battlefield, though a somewhat one-sided one in favour of "offensive" systems. The ballistic missile is analogous with the bomber of air power doctrine, though countering ballistic missiles have so far proven somewhat more difficult than countering the bomber. Anti-ballistic missile weapons that engage targets during the midcourse (in space) phase[51] of a ballistic missile's flight only add further to the argument that space is already a battleground and hence weaponized.[52]

Non-ballistic sub-orbital weapons delivery platforms, proposed as early as the Second World War, also fall into this contentious category. Eugene Sänger's Antipodal Bomber (offered as a means for Nazi Germany to attack the continental US) is perhaps the most well known of such early "space bomber" proposals.[53] Hypersonic weapons are still being proposed for specific demanding roles. Today as before, these proposals rely on speed and altitude to strike deep into an adversary's area of control, bypassing all but the highest reaching defences. However, such aerospace craft are rarely labelled as space weapons, and consequently face less controversy. More attention and controversy seems attached to military spaceplanes (that can actually achieve orbit). Ironically, conventionally armed ICBM's, have been suggested as being more cost-effective and less provocative munitions delivery method than the various military space plane concepts.[54]

The flexible political nature of what defines a space weapons system is highlighted by the academic discussion over whether the Soviet era Fractional Orbit Bombardment System (FOBS) contravened the 1967 Outer Space Treaty (OST), as orbiting Weapons of Mass Destruction are singled out specifically in Article 4 of the OST.[55] FOBS lofted nuclear warheads into an orbital trajectory with the intention of warhead re-entery prior to the completion of one orbit. It was classified by Soviet sources as an ICBM specifically exempt from being described as a space weapons system.[56]

Clearly space weapons do not necessarily have to be stationed in space. Ranging from ASAT weapons, which are easily labelled space weapons, to the various ballistic missile defence (BMD) schemes (which exist with a controversy of their own) ground basing of weapons meant to operate in space is often proposed for the near-term. With the exception of ad hoc weaponization by commanding one spacecraft to ram another (along similar lines to the 1997 Mir supply craft docking accident, but in a more destructive manner),[57] currently available latent forms of space weapons are all terrestrially based. These include existing nuclear armed ballistic missiles set to detonate while in space, along with the already mentioned space warfare laser experiments and ground based BMD systems.[58] In this regard, attacking an orbiting spacecraft would seem similar to attacking a very high flying aircraft.

Basing within the confines of the Earth's atmosphere is attractive technically in that such a system only has to deal with all the rigours of space during the end-game of its attack. As with many launch concepts that try to keep the energy sources firmly fixed to the ground, the infrastructure needed to support such a weapon would not require as demanding engineering as spacecraft. The Mid-Infrared Advanced Chemical Laser (MIRACL) laser facility, which is sometimes labelled as an existing US space weapon, has greatly relaxed criteria for volume and weight compared to the proposed SBL orbiting laser weapon system. In addition, the MIRACL facility in White Sands, New Mexico has been operating for some time now.[59]

Ground basing would not be able to take advantage of the vantage point of space. For many military thinkers, not seeking an elevated position above ones foes would seem irrational. Orbital space then, by virtue of being above all terrestrial targets, would then seem an ideal place for stationing many weapons concepts (if one could only minimize the political concerns and the problem of getting and functioning in space). Unlike ground based systems an orbit based weapon would not have to fight its way against gravity to get to space, it would already be there. George and Meredith Friedman state that

> Ground-based interceptor missiles will have limited zones of engagement – and they will be fighting gravity. Space-based systems, either in low earth orbit or geostationary orbit, will have much wider zones of engagement – and will have gravity as an assist.[60]

A constellation of satellites is needed for an orbit based weapon to be effective. Low Earth Orbit satellites complete several circuits around the Earth per day, meaning no one satellite can provide continuous coverage. Global coverage requirements are demonstrated by the Navstar GPS constellation which has 24 satellites spaced at intervals along six orbital planes.[61] Similar sized constellations are required for effective coverage by multi-shot weapons such as SBL.[62] Brilliant Pebble type systems, as mentioned before, required the maintenance of hundreds of small kinetic energy interceptors each orbiting like a satellite until needed. For either system to be effective against ballistic missile attack emerging from any point on the globe, such infrastructure must be in place prior to open hostilities. An effective constellation would be more or less an all or nothing proposition.[63] Depending on one's perspective on the matter, being continually on station for war in any part of the world, as GPS is today, and as these weapons concepts promise, is either a reassurance of swift defence, or a continual threat.

The exploitation of an advantageous position such as in orbit is technically a sound proposition; however, politically it could be interpreted as a sign of hostile intent. During much of the Cold War, aggressive posturing such as continuous bomber patrols warned the Soviet Union of US resolve in nuclear deterrence. At present there does not seem to be similar rationale for such posturing. Surface to space capabilities have historically seemed less politically aggressive. The limitations imposed by ground based systems would mitigate the impression that defensive systems had more offensive roles in mind. A ground based capability could be portrayed as a reserve or holdout capability for suitably dire situations, much like nuclear weapons.

The political acceptability of ground based space weapons must be balanced against inferior performance, and more difficult targeting solutions. The shift in emphasis from an orbit based Kinetic Energy BMD system (Brilliant Pebble and later Global Protection Against Limited Strikes) to the current ground based National Missile Defence has resulted in a system that, "may be the most politically acceptable but is arguably the least effective and most technologically challenging for defense systems."[64] Among other things, switching from a Brilliant Pebble type BMD to ground based interceptors traded the problems of keeping a mass constellation of weapons operational in space, for the problems of separating warheads from decoys during midcourse phase (space portion) of the target missile's flight.

Boost phase interception (when the target ICBM is still firing its engines and accelerating) is often thought of as being easier than midcourse interception, due to the fact the ICBM is a single target verses the multiple targets presented by an ICBM that has released decoys and potentially multiple warheads or sub-munitions. A ground based BMD system would have to chase down its target. The "tail chase," portion of the interception would limit the time a boost-phase interception could take place. An orbiting boost phase BMD system based on a sufficiently sized and well arranged constellation would already be ahead of all of its potentials targets, surface based ballistic missiles. Orbital basing may also allow for attempts at interception during midcourse and even last ditch attempts against the warheads, though for the above mentioned reasons, attack during the boost phase is clearly preferable from a technical standpoint.

Near term developments may allow more military related space missions to be performed more efficiently and cheaply by orbiting platforms. NFIRE sought to place a missile interception sensor package in orbit, eliminating the need for a BMD interceptor launch for each set of tests (only target missiles needed to be launched).[65] Only slightly smaller than the NFIRE "kill vehicle," micro-satellites (satellites small enough to be built by university students[66]) are often proposed as likely cost efficient methods of accomplishing close observations on target satellites. Boeing's XSS Micro-satellite program aims for the "[d]emonstration of a 25kg autonomous space system in a mission application for Air Force inspection needs."[67]

Taking advantage of being in orbit for the relatively benign mission of being an observer or spotter brings up weaponization as a next step. A satellite that can approach for inspection is certainly close enough to attack. Like the Soviet era co-orbital ASAT weapon or space mine, a weaponized micro-satellite could leisurely adjust its orbit until it was near enough to its target to destroy it.[68] A micro-satellite capable of autonomously flying in formation with a non-cooperative (unaware) satellite would make an ideal space mine. Aside from stealthy approach and inspection, a cheaply produced and launched micro-satellite could, on command, interfere or destroy a target satellite. The method of attack could be as simple as the micro-satellite station keeping in front of the target's camera lens.

For critics, the development of a weaponized micro-satellite would lead to a standoff situation where critical Western satellites are trailed by hostile space mines, which are themselves trailed by Western space mines acting as bodyguards. All these weapons would be potentially part of a pre-emptive strike either to knock out critical infrastructure or to destroy their potential attackers prior to hostilities on the ground. The attacking nation could, if it chose to, simultaneously activate its micro-satellites giving only seconds warning at most. Unlike the ground based co-orbital ASAT which could intercept after a handful of orbits,[69] these micro-satellite space mines would take time to be pre-positioned. As one can imagine most sides of the debate would have an interest in the creation of international guidelines for "safe" separation distances between all controlled spacecraft, if not to prevent the use of space mines then to provide justification for the use of "hard-kill" defences against potential space mines.

The unease over these military observation and remote sensing missions highlights the often mentioned "taboo" against space weapons. A weapon cannot function unless it can "see" its target; hence, it would appear that improving the military's ability to "see" in general would draw criticism over its weaponization potential. In addition to valid technical, financial and strategic impediments, the fact that space has not been weaponized (in the conventional sense) brings special pause and concern. Military advantage, politically, can be a liability if it is perceived as foreshadowing aggressive intent. Without a clearly defined threat, being first to weaponize space may be regarded by some as a risky gamble, or an unnecessary provocation to peers and near-peers.

As for perceptions on space weapons, space as a "battleground" does not necessarily imply the use of the most horrible weapons available. There would certainly be political costs associated with being the first to openly deploy what popular opinion would consider a space weapon. However, in light of the fact that only conventional capabilities are being proposed, the taboo associated would still be outweighed by that of Weapons of Mass Destruction (WMD). There is not quite the same repugnance associated with space weapons as there are with chemical or biological weapons. The fear and loathing reserved for these two specific means of attack in many cases goes beyond that held for nuclear weapons which arguably are more effective at causing mass damage. Space weapons are not by definition weapons of mass indiscriminate overkill.

The only way for a space weapon to acquire the same kind of repugnance as WMD's is to actually fit these weapons with WMD warheads. The strategic nuclear mission from space has not been forwarded as a serious proposal in several decades (aside from critics who believe that once there is precedent for weaponization, orbiting nuclear weapons are an inevitable follow-on). For various operational reasons, including vulnerability (submarines, hardened missile silos and rotating bomber patrols are more survivable than fragile predictable space platforms), control issues, and sheer expense, space based strategic weapons are unlikely to be wanted by even the most ardent proponent of space weaponization.[70] It would appear that the banning of WMD's in orbit, as part of the Outer Space Treaty of 1967,[71] had more utility for public relations than actually pursuing this as a military capability (not withstanding arguments that Soviet FOBS were a demonstration of the utility in breaking the treaty).

Stepping back from the easiest of space weapon systems to identify, those that cause spectacular destruction, there are certainly other less active means of applying coercive force in and from space. Warfare in Clausewitz's words is to "compel our enemy to do our will."[72] Not all means to "compel," in and from space have to be quite as active as hurtling destructive energy and mass at targets.

Perhaps the most benign method to "compel" an outcome out of a space system is to monopolize its services. Paying for exclusive rights to a space service, as was done by the US contracting *Space Imaging* for all overhead imagery of the Afghanistan war zone of 2001-2002 would seem to be effective under the right circumstances.[73] While exclusive access rights do go against the Outer Space Treaty's lofty ideal of equal distribution of benefit to all humanity (there is no mention of belligerents with regard to this point), this type of space control is largely overshadowed by the more controversial issue of space weapons.

Passive means such as simply "getting-in-the way" of a satellite's operation, but not actually harming it, may even be the only feasible method for a space power to "attack' objects in space without directly jeopardizing its own space assets through the creation of additional space junk. The remains from attacking a spacecraft in orbit, unless carefully planned to cause immediate de-orbit, will tend to continue circling the Earth and become a hazard to friendly space assets. Due to the lasting consequences of debris in orbit, neutralization of targets in space via traditional means of grievous and spectacular physical destruction may not even be desired. Many space weapons currently being proposed have more in common with so called "less-than-lethal" weapons.

As mentioned earlier, the US officially wants a range of effects from degradation to permanent neutralization of the services being provided from space. Most of these effects do not require physical damage to the target spacecraft itself. This desire for a range of effects can be found in the doctrine of Flexible Negation. The concept of Flexible Negation "involves such measures as jamming, spoofing, and blinding enemy satellites and disabling enemy ground support stations."[74] These are commonly regarded as being passive means. No permanent "coercive" force is expended against a target in space or from space. According to Benjamin Lambeth,

> As for follow-on measures toward acquiring a more active space control capability, those who would seek the benefits of such a capability without transgressing the taboo against migrating armed combat into space have proposed a mode of operations called "flexible negation" in lieu of direct attack.[75]

Oddly enough, "less-than-lethal" capabilities have come under criticism for the ability to tone down damage and avoid permanent effects. Such "Public Relations friendly" consequences, it is argued, would tend to invite their use more often (usually in the context of suppressing "legitimate protests").[76] Along the same lines, the fact that the satellites of present military concern are generally not crewed has led to the suggestion that attacking them would be a more humane method of war.[77]

Conceivably, an argument could be built around this lack of direct human casualties in space warfare as being encouragement to bypass diplomatic alternatives and go straight to the military options. From here escalation is certainly feared (where there would be direct human casualties). This fits into the general argument that desensitization to violence (separation of consequences from actions) leads to more violence, and is not unique to the space weaponization debate.

Though not generally thought of as a weapon, electronic warfare would be a relatively simple way to "negate" space assets. If destruction of satellites is considered a politically sensitive topic, then simply denying their use by an opponent would be a reasonable relatively uncontroversial substitute. Already there is at least one electronic warfare system specifically targeting satellite communications being procured by the United States. In September 2004, the Counter Communications System (CounterCom), a ground based satellite communication jamming system, was declared operational by the USAF.[78]

Satellites represent bottlenecks or chokepoints in the flow of information. This then leads to the argument favouring the ability to negate satellites in space. As satellites in orbit are in constant movement relative to points on Earth (with the exception of GEO), electronic warfare being conducted against the satellite as opposed to its customer base on the ground would require the placement of jamming equipment on spacecraft capable of on-orbit rendezvous with target satellites. While conceptually this would be analogous to electronic warfare aircraft inhibiting enemy sensors and communications, such a capability despite being reversible and benign would constitute a "soft-kill" ASAT weapon. Indonesia's Palapa B1, as mentioned earlier, is accused of performing this kind of service denial when its broadcasts interfered with those of a satellite owned by a Hong Kong based company from getting through to its subscribers in the People's Republic of China.[79] As the satellite operators were in conflict over who had use of a particular orbital slot, Palapa B1's activities could be regarded as an intentional hostile act in space. According to one account, the broadcasts were justified by Indonesia with claims that it was not treaty bound to avoid the broadcast of those interfering signals.[80]

More clearly defined space weapons would be those that cause permanent damage to targets. As on Earth, target destruction is usually accomplished by the delivery of energy from weapons platform to target. Electronic warfare techniques, normally thought of as being non-destructive can cause permanent damage if the energy level being broadcast is high enough. High power microwave (HPM) and other radio frequency (RF) weapons are being investigated for their ability to tune effects from temporary disablement to destruction of fragile electronic components.[81] These types of RF systems would go beyond simple jamming, and into the territory of Directed Energy Weapons (DEW).

The laser is a more commonly thought of DEW system, and certainly has science fiction overtones. Despite the sometimes aggressive image laser technology holds in the general public's mind, laser attacks do not necessarily have to be permanent. Like the RF base jammer/weapon concept, there is potential for tuning the effects of a laser to temporarily blind optical sensors only. While blinded, an optical reconnaissance satellite has as much value to its operators as one that was shattered by a physical attack.

There are many ways a satellite may "attack" or be "attacked." Aside from imagination and paranoia, orbiting would be the critical characteristic for systems perceived to be space weapons. During the Cold War, even the very ability to interact routinely with other orbiting spacecraft was viewed as threatening. Canada's Canadarm, also known as the Shuttle Remote Manipulator System (SRMS), was very much a factor in these concerns. "Concern has been expressed by the Soviet Union about the shuttle's ability to pluck satellites from orbit."[82] Present day criticism over NFIRE and microsatellites reinforces the political definition of the space weaponization benchmark as routinely having weapons that are stationed in orbit and/or weapons that are specifically for attacking targets in orbit.

Spaceflight in general becoming more robust, losing much of its experimental quality and gaining an industrial mode of operation, would fulfill all of the technical prerequisites for space weaponization as it is commonly perceived. However, having the technology, opportunities, and even vulnerabilities are not enough to make weaponization a certainty. While it is useful to draw analogies to the weaponization of aerial and maritime mediums, the militarization (or seeming lack) of the Internet medium shows that military developments must be tailored for what can be done in the medium.

Many have described the military use of space as a continuum or linear progression.[83] Lupton's Four Military Space Doctrines on such an order would list: sanctuary, survivability, space control and finally high-ground.[84] These correspond with: no weapons in space, being able to survive an attack on critical space assets, space as only one important battlefield among many (analogous to sea and air control), and space weapons allowing for dominance not just over tactical battlefields, but over strategic outcomes. Of Lupton's four, only the latter two demands a US space weapons capability (though the second requires that the technology necessary for an attack to have proliferated to powers hostile to the US). Indeed, except for the more aggressive types of force application in and from space (some space control and most high ground concepts) active measures may even be unnecessary.

Generically, intelligence gathering, militarily useful Earth measurement, and communications are usually found on the low end of the many space militarization spectrums. This corresponds to the early use of space by the military. The present integration of terrestrial weapons with space systems is found in the middle of these continuums. Finally at the top or end of these attempts to chart the progress of space militarization, the various options concerning the development and use of weapons in space are listed. The highest "threat" level usually found on such descriptions of space militarization, are the space to surface force projection capabilities.

Grouping military use of space along a timeline clearly demonstrates the limiting nature of technology. The requirements for basic Earth observation were achieved in the early decades of the Space Age, first for the grand strategic picture, and later down to the tactical level. Later developments allowed individual platforms and weapons to have integrated satellite support, forming a space dependent "recon-strike complex." While space weaponization options beyond today's latent capabilities are only speculation, their expected development is listed in order of technological difficulty as well. Starting with ground based, or air-to-space weapons, these are logical progressions from existing "latent" capabilities found in the MIRACL facility[85] and developments of systems originally intended for the missile defence role. Ballistic missile defence systems have, for similar technical and political reasons,[86] started with ground based systems. Space based (active and passive) systems are proposed as advanced follow-on capabilities.

A significant, non-nuclear, orbit to surface capability is usually regarded as the most technologically challenging of space weaponization roles. The greatest challenge would seem to be getting the weapon (or weapon effect in the case of a DEW system) to enter the atmosphere with precision. For kinetic energy bombardment schemes, precision is critical in that such weapons have only their own (tremendous) force of impact as the destructive mechanism. Slowing down a weapon may be necessary to allow it enough time to manoeuvre before impact, but this would also lower the kinetic energy involved. Below a certain speed threshold a space based strike system will have little to offer compared to an air or surface launched strike alternative aside from global presence. Significant technological advances will have to be made before such a "bolt-from-the-blue" capability is possible, let alone useful.

At the same time a space to surface capability is the most troublesome capability to space weapons critics on several counts; it is the most expensive capability (possibly being redundant to capabilities offered by future system operating totally within the atmosphere), and is seen by many as being unnecessarily provocative. An orbiting weapon once deployed has coverage on all targets below its orbit track within its limits of manoeuvrability. Once in orbit such a strike weapon would give its target little warning, it is always ready to fall from the sky without warning.

Omnipresent global presence with little to no warning of attack would be the space version of having foreign bombers continuously overhead or warships sitting always within strike range. In other words it would be similar to the continual threat of ICBM attack during the Cold War, but without the involvement of nuclear warheads and their deterrent effect.[87] This logic however may be applied to any of the future rapid global strike options the US is pursuing.

As for hypothetical conventional orbit-to-surface strike weapons, the same charge of being a "first strike weapon" was made for the nuclear armed FOBS. From launch pads in the Soviet Union, FOBS could place a nuclear weapons payload into an orbital track that avoided US early warning radars.[88] Such a weapon could potentially accomplish a successful first strike on US nuclear forces and therefore avoid retaliation. Here space related technology, if not an actual space to surface offensive weapons system, was offered to fulfil a specific role in Soviet nuclear strategy. Ominously, perhaps, space will once again offer a potential foe a viable military solution to counter US capabilities.

Additionally critics argue that an orbit based strike system would be destabilizing in that such a system would share the vulnerabilities of all contemporary satellites, and therefore their use would be encouraged for fear of losing such systems in a potential attack. This of course includes the Cold War fear of accidental firing due to "tripwire" posturing, though how far this may be applied to a conventional strike weapon is debatable.

This type of criticism of the general idea of placing weapons in space overlooks the fact that an orbital strike system is not the only global range conventional strike option available to the US. The great difficulties (both technical and cost) that appear to be inherent in space weapons certainly dampens enthusiasm for orbit to surface strike systems. Though the vantage point and opportunities of space are impressive, it may not be necessary or even desirable to take advantage of it for every mission type. Quite possibly, orbital weapons may be of a purely defensive nature, and therefore avoiding the provocative aspect of being a strike weapon.[89]

Nor would space weapons be the only manifestation of intentions to deny an adversary warning of attack. Today, stealth aircraft and cruise missiles do not have the stigma of being "first-strike weapons," despite their well proven use on the "first day of war" to incapacitate early warning, defences, and communications, command, control and intelligence (C3I) infrastructure. Space to surface systems, like bombers, long range naval artillery and ballistic missiles of various ranges (which conceptually are all often in direct competition for the strike mission) all have immediate application as offensive weapons. Due to the inherent cost factor of space systems, a space to surface strike capability would have to be justified by the existence of targets extraordinarily resilient to other means of attack.

Concerns exist that once space becomes weaponized for roles labelled as being "defensive," it follows that "offensive" missions only become inevitable. This however overlooks the argument that space weaponization has already occurred with the "offensive" strike role first, in the form of ballistic missiles. Indeed, from the use of the term "first strike" in the criticism of space weaponization there appears to be some belief that nuclear (de)stability arguments would be immediately applicable to what are currently being proposed as conventional warfare systems.

The early history of practical spaceflight and the nuclear standoff of the "Cold War," are certainly interlinked. Some have floated the idea that spaceflight's association with nuclear technology has given rise to much of the caution and concern over weaponizing spaceflight and space, "possibly because in its developmental stage, as a result of space flight's direct association with ballistic missile and nuclear weapon development, a chord of universal terror was struck in our communal consciousness."[90]

During the Cold War, policies regarding weapons in space were clearly guided by what the technology and the relationship between the superpowers allowed. To fully consider the potential for continued space militarization, and whether it will lead to space weaponization it is necessary to consider the background of politics and technology.

Notes

1. Fédération Aéronautique Internationale, "100 km Boundary for Astronautics," 25 June 2004. <http://www.fai.org/book/view/22> (2004).

2. British Broadcasting Corporation, "SpaceShipOne rockets to success," 4 October 2004. <http://news.bbc.co.uk/1/hi/sci/tech/3712998.stm> (2004).

3. Falling is described as gravity pulling two objects together, or the perception of the smaller mass being drawn "down" to the much, much greater mass.

4. National Aeronautics and Space Administration, "Orbital Velocity and Period Calculator," 23 June 1995. <http://liftoff.msfc.nasa.gov/academy/space/atmosphere.html> (2005).

5. National Aeronautics and Space Administration, "Skylab Operations Summary," 29 September 2000. <http://www-pao.ksc.nasa.gov/kscpao/history/skylab/skylab-operations.htm> (2004).

6. National Aeronautics and Space Administration, "Earth's Atmosphere," 1 December 1995. <http://liftoff.msfc.nasa.gov/academy/space/atmosphere.html> (2004).

7. Lighter-Than-Air Craft (balloons and airships) are being proposed as platforms for operation at extremely high-altitudes with long on-station endurance. Such capabilities may supplant satellites in some roles.

8. National Aeronautics and Space Administration, "Earth's Atmosphere."

9. Greg Klerkx, *Lost in Space: The Fall of NASA and The Dream of a New Space Age*, (New York: Pantheon Books, 2004), 65.

10. That is not to say that the several scores of kilometers between space and air capable of supporting powered flight are unimportant. Sustained operation in this region by specialized lighter than air-craft will be discussed later as a possible alternative to many space systems.

11. The University of Tennessee, "Newtonian Gravitation and the Laws of Kepler," 4 October 2004. <http://csep10.phys.utk.edu/astr161/lect/history/newtonkepler.html> (2004).

12. Kenneth W. Barker, "Airborne and Space-Based Lasers," in *The Technological Arsenal*, ed. William C. Martel (Washington: Smithsonian Institution Press, 2001), 45.

13. Prominent examples of in space repair and construction are the Hubble Space Telescope and International Space Station programs respectively. Both programs have had difficult histories.

14. Going up only requires that one is able to generate an acceleration to overcome that imparted by gravity on all objects. At sea level the acceleration between the Earth and any object is merely 9.8 m/s2 (metres per seconds squared).

15. Tim McElyea, A Vision of Future Space Transportation, (Burlington: Apogee Books, 2003), 48.

16. Andrew J. Butrica, Single Stage to Orbit, (Baltimore: The Johns Hopkins University Press, 2003), 70.

17. British Broadcasting Corporation, "Brazil vows to pursue space plan," 23 August 2003. <http://news.bbc.co.uk/1/hi/world/americas/3176395.stm> (2004).

18 Greg Klerkx, *Lost in Space: The Fall of NASA and The Dream of a New Space Age*, 95.

19 Ibid, 132.

20 James Oberg, "The war of words over war in space," 16 April 2004. <http://msnbc.msn.com/id/4732874> (2004).

21 One of NFIRE's goals is to observe intercontinental ballistic missile (ICBM) flight characteristics to better refine interception techniques. The hardware actually being put into orbit to monitor missiles in flight is a ballistic missile interceptor (labelled as a kill vehicle) minus the propulsion needed for major manoeuvres.

22 Ibid.

23 Greg Klerkx, *Lost in Space: The Fall of NASA and The Dream of a New Space Age*, 38.

24 Daniel G. Dupont, "Nuclear Explosions in Orbit," Scientific American 290 no. 6 (July 2004): 105.

25 Ibid, 103.

26 Ibid.

27 Geosynchronous orbits bear the name of Clarke orbits after Arthur C. Clarke, of Science Fiction fame, who first proposed the idea in 1945. Due to a fine balance between extremely high altitude and gravity, a spacecraft in a GEO orbit matches its orbital speed with the speed of Earth's rotation, causing it to appear fixed above a particular point on the surface. As the "fixed" effect of GEO can only occur in a narrow band around the equator, there is certainly a limit as to how many interests can benefit from such strategically important positions.

28 Everett C. Dolman, *Astropolitik*. (Portland: Frank Cass, 2002), 134.

29 Though it is generally the crewed spacecraft that makes the effort to get out of the way of an object that has left the control of humanity and is now subject to the Laws of Physics.

30 The Eisenhower Institute, "A European Perspective on Current Trends in Military and Civilian Space," 2004. <http://www.eisenhowerinstitute.org/programs/globalpartnerships/fos/newfrontier/parismeeting.htm> (2005).

31 Ibid.

32 It is of note that the satellite being jammed was attempting at the time to supply commercial services to the People's Republic of China and was owned by APT Satellite Company, based in the then British colony of Hong Kong. The satellite that "challenged" Indonesian occupation of the disputed slot in 1992 was owned by US based, Rimsat.

33 Simon P. Worden, "Space Control in the 21st Century: A Space 'Navy' Protecting the Commercial Basis of America's Wealth," in Peter L. Hays, James M. Smith, Alan R. Van Tassel, and Guy M. Walsh, eds., *Spacepower for a New Millennium* (New York: McGraw-Hill, 2000), 227.

34 This is notwithstanding the limited militarization (or other exploitation) of the sea bed at present time. Deep ocean depths, well below that which most military submarines safely operate at, present an environment that is at the very least as demanding to operate in as space.

35 The Commission to Assess United States National Security Space Management and Organization, *Report of the Commission to Assess United States National Security Space Management and Organization*, 11 January 2001. <http://www.space.gov/docs/fullreport.pdf> (2004).

36 Michael E. O'Hanlon, *Neither Star Wars Nor Sanctuary*, 8.

37 United States Air Force. *Air Force Doctrine Document* 2-2.1, 2 August 2004. <http://www.dtic.mil/doctrine/jel/service_pubs/afdd2_2_1.pdf> (2004).

38 United States Strategic Command. *Space Missions*, March 2004. <http://www.stratcom.mil/factsheetshtml/spacemissions.htm> (2004).

39 Max Boot, "The New American Way of War," *Foreign Affairs* 82, no 4 (July/August 2003). <http://www.foreignaffairs.org/20030701faessay15404/max-boot/the-new-american-way-of-war.html> (2004).

40 Global Security. "BGM-109 Tomahawk," <http://www.globalsecurity.org/military/systems/munitions/bgm-109-var.htm> (2004).

41 Ibid.

42 Strategic nuclear weapons such as in service Intercontinental and Submarine Launched Ballistic Missiles (ICBMs and SLBMs) still rely on inertial guidance methods, with a nuclear warhead "smart bomb" like accuracy is unnecessary. There is however nothing technically challenging about using satellite assisted guidance on strategic nuclear weapons if the need ever arose.

43 Jeffrey McKitrick, et al., "The Revolution in Military Affairs," *Battlefield of the Future*. <http://www.airpower.maxwell.af.mil/airchronicles/battle/ov-4.html> (2004).

44 Department of State, "Treaty on Principles Governing the Activities of States in the Exploration and Use of Outer Space, Including the Moon and Other Celestial Bodies," 27 January 1967. <http://www.state.gov/t/ac/trt/5181.htm> (2004).

45 Conversely, space systems being integral to tactical weapon systems has shaped the RMA described as ongoing today.

46 This however does not stop a soldier from calling down fire on coordinates occupied by friendly forces, civilians and other non-combatants, or even their own location mistakenly.

47 George Friedman, and Meredith Friedman, *The Future of War*, (New York: St. Martin's Griffin, 1996) 270.

48 Karl P Mueller, "Totem and Taboo: Depolarizing the Space Weaponization Debate," (Paper based on presentation given to Weaponization of Space Project of the Eliot School of International Affairs Space Policy Institute and Security Policy Studies Program, George Washington University, 3 December 2001); available at <http://www.gwu.edu/~spi/spaceforum/TotemandTabooGWUpaperRevised%5B1%5D.pdf> (2004).

49 The Commission to Assess United States National Security Space Management and Organization, *Report of The Commission to Assess United States National Security Space Management and Organization*, 11 January 2001. <http://www.space.gov/docs/fullreport.pdf> (2004).

50 John M. Logsdon, "Just Say Wait to Space Power," *Issues In Science and*

Technology 17, no 3 (Spring 2001).
<http://www.issues.org/17.3/p_logsdon.htm> (2004).

51 This would include the now being fielded US Ground Based Interceptors (GBI) of National Missile Defense (NMD), and also any tactical ballistic missile defence that was capable of intercepting a missile in space during the midcourse phase of ballistic missile flight.

52 By this definition of space weaponization the continuous deployment of exo-atmospheric ABM's (such as the early Galosh system) protecting Moscow have made space a potential battlefield for a few decades.

53 Robert Godwin, ed. *Dyna-Soar Hypersonic Strategic Weapons System.* (Burlington: Apogee Books, 2003), 7.

54 Bruce M. DeBlois, Richard L. Garwin, R. Scott Kemp, and Jeremy C. Marwell. "Space Weapons: Crossing the U.S. Rubicon," *International Security*, Fall 2004.
<http://mitpress.mit.edu/catalog/item/default.asp?ttype=4&tid=26> (2005).

55 Department of State, "Treaty on Principles Governing the Activities of States in the Exploration and Use of Outer Space, Including the Moon and Other Celestial Bodies," 27 January 1967.
<http://www.state.gov/t/ac/trt/5181.htm> (2004).

56 Hung Nguyen, "Russia's Continuing Work on Space Forces," Orbits, Summer 1994, 413-423, quoted in Matthew Mowthorpe, *The Militarization and Weaponization of Space,* (Toronto: Lexington Books, 2004) 70.

57 British Broadcasting Corporation, "Mir Space Station 1986-2001," <http://news.bbc.co.uk/hi/english/static/in_depth/sci_tech/2001/mir/default.stm> (2004).

58 Michael E. O'Hanlon, *Neither Star Wars Nor Sanctuary*, (Washington, DC: Brookings, Institution Press, 2004), 27.

59 David Hobbs, *Space Warfare*, (New York: Prentice Hall, 1986), 110.

60 George Friedman, and Meredith Friedman. *The Future of War.* 342.

61 Global Security. "Navstar Global Positioning System," <http://www.globalsecurity.org/space/systems/gps.htm> (2005).

62 Matthew Mowthorpe, *The Militarization and Weaponization of Space,* (Toronto: Lexington Books, 2004) 156.

63 An incomplete constellation would allow for unhindered ballistic missile attack during periods without coverage. Knowing when such periods would occur would tend to be easier with large easy to observe multi-shot satellites than with hundreds of independent interceptors. Then again, it is arguable that incomplete knowledge interception coverage alone could hamper the ICBM plans of smaller nations.

64 Peter Hays, et al, ed, *Space Power for a New Millennium*, 15

65 James Oberg, "The war of words over war in space."

66 Christopher A. Kitts and Richard A. Lu, "The Stanford SQUIRT Micro Satellite Program," 7 June 7 1994.
<http://ssdl.stanford.edu/aa/papers/SSDL9404.pdf> (2004).

67 Boeing, "XSS Micro-satellite," <http://www.boeing.com/defense-space/space/xss/> (2004).

68 Matthew Mowthorpe, *The Militarization and Weaponization of Space*, 117.
69 Ibid, 123.
70 Karl P Mueller, "Totem and Taboo: Depolarizing the Space Weaponization Debate," (Paper based on presentation given to Weaponization of Space Project of the Eliot School of International Affairs Space Policy Institute and Security Policy Studies Program, George Washington University, 3 December 2001).
<http://www.gwu.edu/~spi/spaceforum/TotemandTabooGWUpaperRevised%5B1%5D.pdf> (2004).
71 Department of State, "Treaty on Principles Governing the Activities of States in the Exploration and Use of Outer Space, Including the Moon and Other Celestial Bodies," 27 January 1967.
<http://www.state.gov/t/ac/trt/5181.htm> (2004).
72 Carl von Clausewitz, *On War*, Everyman's Library Ed. trans. and ed. Michael Howard and Peter Paret (Princeton University Press, 1976 ; reprint Toronto: Alfred A. Knopf, 1993), 83.
73 "Eye spy." *The Economist*, Nov 10, 2001, Quoted in Global Security, <http://www.globalsecurity.org/org/news/2001/011110-eye.htm> (2004).
74 Benjamin S.Lambeth, *Mastering the Ultimate High Ground: Next Steps in the Military Uses of Space*, RAND,
<http://www.rand.org/publications/MR/MR1649/MR1649.ch5.pdf> (2004).
75 Ibid.
76 Lev Grossman, "Beyond the Rubber Bullet," *Time Online Edition* 21 July 2002. <http://www.time.com/time/nation/article/0,8599,322588,00.html> (2004).
77 Michael E. O'Hanlon, *Neither Star Wars Nor Sanctuary*, (Washington, DC: Brookings, Institution Press, 2004), 18.
78 Jeremy Singer, "Satellite Jammer Ready: U.S. Parallel Effort To Thwart Imaging Craft Dropped," C4ISR Journal 19 October 2004.
<http://www.c4isrjournal.com/story.php?F=461040> (2004).
79 The Eisenhower Institute, "A European Perspective on Current Trends in Military and Civilian Space," 2004.
<http://www.eisenhowerinstitute.org/programs/globalpartnerships/fos/newfrontier/parismeeting.htm> (2005).
80 Ibid.
81 Eileen M. Walling, "High-Power Microwaves and Modern Warfare," in *The Technological Arsenal*, ed. William C. Martel (Washington: Smithsonian Institution Press, 2001), 94.
82 David Hobbs, *Space Warfare*, (New York: Prentice Hall, 1986), 145.
83 Lt. Col., USAF, Bruce M. Deblois, "Space Sanctuary A Viable National Strategy," *Aerospace Power Journal*, Winter 1998.
<http://www.airpower.maxwell.af.mil/airchronicles/apj/apj98/win98/deblois.html> (2004).
84 Lt. Col (Retired)., USAF, David E. Lupton, *On Space Warfare*, (Maxwell Air Force Base, Alabama: Air University Press, 1998).
<http://www.airpower.maxwell.af.mil/airchronicles/apj/apj98/win98/deblois.html> (2004).

85 Michael E. O'Hanlon, *Neither Star Wars Nor Sanctuary*, 27.

86 Peter L. Hays, James M. Smith, Alan R. Van Tassel, and Guy M. Walsh, *Space Power for a New Millennium*, (New York: McGraw Hill, 2000), 14

87 This comparison once again highlights the ironic suggestion that conventionally armed ICBMs would be less provocative as a global strike system than an orbit based system.

88 A traditional ICBM flight profile has the rocket engine(s) pushing the payload up instead of generating the great momentum needed to keep the payload in orbit. While less energy is involved in a ballistic launched compared to an orbital launch, long range ICBM's do have to arc high above the surface to reach their targets. An orbital path does not present as much exposure over the horizon for detection, and can approach from any direction as it is not limited by range.

89 On the other hand, it is argued by advocates of pure deterrence stances that a perceived effective defence is destabilizing in that the side under such a defence would believe themselves able to escape retaliation, and therefore make it likely for it to strike first. Similarly in preventing one side from achieving a perceived effective defence, the other may decide on preemptive action prior to the defence's completion.

90 Everett C. Dolman, *Astropolitik*. (Portland: Frank Cass, 2002), 169.

CHAPTER TWO

Military Space Potentials

Information and Sanctuary

Cold War fears, like many fears and misunderstandings in history were affected by a lack of information. Lack of information regarding a competitor's intentions and capabilities, are a source of insecurity. Cold War fears over "gaps" between Soviet and US capabilities drove many costly defence projects. After exhausting often sparse open sources, the usual means of acquiring information, espionage, is usually regarded as an unfriendly act, often raising tensions further. The aftermath of Francis Gary Powers' U2 being shot down, the political embarrassment to the US and hardship endured by pilot Powers after his capture, amply demonstrated the dangers overflights entailed. Yet without this type of information, nations would be left fumbling around in their relations, perhaps risking war. In a nuclear armed world, such a situation is far from ideal.

Observing from space does not carry the same risks associated with terrestrial espionage. As mentioned before, it would be difficult to control the space overhead without a robust space warfare capability. Though not without protests there eventually seemed to have developed during the Cold War an acceptance that satellite intelligence gathering was a stabilizing influence. Indeed there is little point in hiding a deterrent force from intelligence satellites, unless these satellites could meaningfully contribute to the deterrent's negation. Limitations on early Cold War era satellite technology (small numbers and long delays in retrieving data from space), allowed satellite reconnaissance agencies to know of the existence of a deterrent force, but prevented enough detail to aid tactically in their negation. Nonetheless as a means of developing a relatively accurate picture of adversarial deterrent capabilities, space reconnaissance satellites had a stabilizing effect on Superpower relations.

Up until the last few years of the Cold War the benefits of this stabilizing effect have outweighed the benefits of developing the means to neutralize such satellites. This led to the view that space was a "sanctuary" for these types of missions during the Eisenhower administration and throughout all administrations up to and including Carter.[1] Though both sides of the Cold War were engaged in the testing of possible ASAT technologies during this period, there was no great push to use the limited spaceflight capabilities each had to negate the few reconnaissance satellites that were active at any time. Many such as Arthur Schlesinger and Philip Klass[2] refer to a "tacit" agreement between the superpowers to avoid direct actions against one another's space assets.

> ...both the United States and the Soviet Union recognized the mutual benefits of reconnaissance satellites and reached a 'tacit' agreement to refrain from developing weapons to counter them.[3]

As demonstrated by the Soviet Union softening its objections to US intelligence activities in outer space, both parties found strategic reconnaissance from space to be mutually beneficial (especially since the Soviet Union's own capability was increasing at the time). Changes to international relations may once again restore space as being the only acceptable realm for the mission of treaty verification.[4] Advances in technology permitting mass deployment of space weapons would, according to the sanctuary position on space power, endanger this mission. That is if space surveillance in this Cold War mode remains of paramount importance.

Information superiority is often touted as being the foundation of security for the Western world in the 21st century. Overblown phrases such as "electronic Pearl Harbor"[5] are used to emphasise the importance of information. Arguments for and against space weapons certainly keep in mind that much of the information necessary for US security is collected and relayed both by commercial and government owned satellites.

Certain advocates of space as a sanctuary view this position as forestalling the development of foreign ASAT capabilities that would threaten existing military infrastructure in space. The development of a robust ASAT capability by any power, even for the explicit purpose of satellite defence, could raise fears that such systems could be used for offensive duties. There may be no practical distinction between the capability to stop an attacking spacecraft and that needed to stop the function of a target satellite that was otherwise militarily troublesome. This in turn may lead nations fearing US intentions to develop their own ASAT capabilities, endangering US satellite infrastructure. In other words, "the United States has the most to lose from premature deployment of space weapons, because we depend on space more than any other nation."[6] A policy of delaying deployment of weapons in space is argued as being a confidence building measure meant to set an example to the rest of the world.[7]

Space systems currently available to the US military already provide unparalleled abilities. It would not be too terribly detrimental to US national security to have all space capabilities for all nations frozen in its present form. Freezing technology for all would also lock in place all of the advantages of satellite services enjoyed by US military forces on the ground. Maintaining the space status quo, including space as a sanctuary for satellite operations protects current US conventional military superiority.

The amount of time the sanctuary stance can be maintained is potentially nearing an end. As demonstrated by the global interest in private spaceflight, aerospace technology, much of it of a dual-use nature, is spreading internationally. It is unclear as to how long before space technology proliferation allows a potential foe the ability to both replicate the services provided by US space systems and to negate the services deemed critical to US operations. The use of the phrase, "premature deployment" is only a warning over the timing of space weaponization, and not a complete denouncement of the concept.

Within the opposition to space weaponization, there are certainly those that do denounce the concept of space weapons as being "evil." Though it is a well worn cliché from space, there appear to be no boundaries, no nations, just one Earth. From this idealistic view of the planet partially stems the idea that space can be free of weapons through cooperation and agreement. The treaties concerning Antarctica's status as a sovereignty-free territory seemingly free from warfare are often proposed as a template for cooperative governance of space. An arms race in space is anathema to this point of view.

This view however overlooks or minimizes the true value of the Antarctic to national elites in the US, or for that matter any other major powers. At present, the disposition of Antarctica is hardly relevant to national security. Indeed the majority of the southern continent's tangible value is for scientific research and associated prestige. Cooperation during the Cold War on Antarctica, as a means of dialogue and confidence building, seems to have outweighed whatever military value that the southern continent possessed.

There is proven strategic utility in denying a foe access to the common territories of the sea and international airspace in times of crisis. Space, as has been mentioned several times already is very important to US defence capabilities, and is integral to a modern economy. Denying space to an adversary would also be useful against symmetrical foes. However based on what is currently known, there is little to be gained from denying access to Antarctica.

Given enough utility, it would be irresponsible for a Great Power to willingly give up a useful capability through treaty (without significant concessions in return). Space weapons are expensive and difficult pieces of technology, but the promise of certain capabilities appears to outweigh any benefits from banning the entire concept of weapons in orbit and/or able to attack satellites in orbit. It should be remembered that while China is proposing space weapons bans,[8] there is also evidence that it is also pursuing these capabilities.[9]

A more recent example of this type of behaviour among nations in the security arena would be that of the vaunted landmine treaty and the fact it does not have universal support. Of interest are that the United States, Russia, China and the two Koreas are not currently signatories to the 1997

Convention on the Prohibition of the Use, Stockpiling, Production and Transfer of Anti-Personnel Mines and on their Destruction.[10] That is not to say that these countries are busy planting landmines around the world. Only that these nations still recognize some potential merit in the use of antipersonnel landmines, and have taken steps at least keep the option (treaty) obligation free if not "taboo-free."

US refusals to endorse similar agreements concerning space weapons would tend to indicate a similar belief in possible future utility. The fear over the proliferation of advanced military space capabilities (short of weaponization), and fears over the vulnerability of US satellites, has forced a re-examination whether "space as sanctuary" will continue to benefit US security. This change of policy potentially leads to other space postures, including space denial and space control.

Space Denial, Survivability, and Space Control

Space denial is equivalent to a sea denial strategy. While an adversary would attack US space capabilities, including those in space, it would not, or could not, seek to take advantage of space for its own operations. An example of a sea denial strategy would be commerce raiding, as in the German U-boat campaigns of the First and Second World Wars. Submarine warfare in the two World Wars was primarily against merchant shipping. U-boats, the tools of German sea denial strategy, were a relatively low cost solution when compared to the battle fleets necessary for a sea dominance or command of the sea strategy. Yet for all the damaged caused by German U-boats in the two world wars, this type of warfare did not give it supremacy over the seas. While a sea denial strategy could over time reduce an opponent's ability to sustain a sea dominance strategy, it does not necessarily imply that the nation employing denial of the sea tactics will inherit dominance for itself.

Similarly, as a counter to airpower, a state may employ batteries of lower cost interceptors, surface to air missiles (SAM), and anti-aircraft artillery (AAA) to stop heavy bombers, and in the process challenging classic command of the air strategy as promoted by Guilio Douhet. Arguments have been made that the steps taken to make the bomber concept more survivable with concepts such as stealth and electronic warfare, have actually made anti-aircraft systems more cost effective, in that the costs to continue performing the heavy bombing mission rise faster than for steps taken to counter it.[11] If the costs of defending satellites outweigh the costs of attacking them, then one could make the same arguments against these satellites. This then leads to arguments against a satellite dependent military and strategy.

Space denial as a strategy, does not require large investment or precision. Blunt weapons may be used to great effect, but with consequences. Following a purely space denial doctrine could actually inhibit a nation's own aspirations to gain control or command of space. Aside from depriving funding from potential space control and/or dominance projects, a space denial doctrine does not have to make allowances for friendly and neutral satellites. Space debris generated by physical attacks, and radiation effects from nuclear detonations in space have a tendency to linger, putting at risk any satellite that passes through the aftermath. For this reason some argue that a space denial strategy would be irrational for the US and near peers; "physical attacks in space have a 'scorched Earth' aspect that makes them unattractive to rational actors."[12]

Then again "scorched Earth" tactics have been used successfully in the past. Though causing great hardship, Wellington's strategy to starve the French in Portugal (1811-1812), and Stalin's scorched Earth tactics against the Germans in the Second World War were successful in blunting invasion. As services provided by satellites in space make up a foundation or "centre-of-gravity," for the US military, then denying such things in turn would be of great utility to a military challenger.

Due to the asymmetrical level of capabilities between the US and the next most capable space-faring nation, rudimentary blunt counter-space technology may be of particular value to a foe. The task of space denial is arguably easier than that of space control. Satellites are very fragile objects that still fall apart early just by being in space. Military satellites, though often hardened beyond those put into space by civilian organizations, remain vulnerable to various manners of attack. It should not be forgotten that civilian satellites are not only important to the world economy, but have been outsourced to take on critical military roles. The vulnerability of US space assets has been often alluded to with the provocative warning of a "Space Pearl Harbor."

> An attack on elements of U.S. space systems during a crisis or conflict should not be considered an improbable act. If the U.S. is to avoid a "Space Pearl Harbor" it needs to take seriously the possibility of an attack on U.S. space systems.[15]

Of all the nations that would suffer from the effects, it is of no question that the US and its allies would lose the most. This very asymmetry is why a concerted attempt to destroy Western (or everyone's) space infrastructure is argued as so tempting. From the perspective of a potential foe it is only logical to attempt to counter the massive advantage technology gives to the US. Without satellite communications and navigation, systems such as long range drones and many precision guided munitions would cease to be effective.

The US would indeed have the most to lose in such a conflict, but at the same time, it must be hesitant in allowing a potential foe the same stake in space that it has. The consequence of giving a foe as much to lose in space would be the loss of the asymmetric advantage possessed by the US. As a diplomatic solution to the balance of power in space capabilities would not seem rational in a realist sense to either the US or a potential rival, military force (and threats of) become the only viable solution aside from hoping for the best.

As mentioned before, high-technology may not even be needed to blunt US space and information superiority. A nation less concerned with satellite fratricide, perhaps due to limited use for satellites and satellite services, can always rely on the latent counter-space abilities of a nuclear warhead set to detonate at high altitude. This brute-force approach is certainly available to many adversaries and competitors of the United States.[16]

Even without the devastating effects of a nuclear warhead, any nation or indeed organization with sub-orbital access can put low orbiting satellites at risk. Littering the space in the immediate path of an orbiting satellite with debris, as one would use caltrops against horse cavalry and wheeled vehicles, is often suggested as an example of how simple it would be to destroy a satellite.[17] A ballistic missile would put satellites orbiting at or below the apex of the missile ballistic flight within reach of this manner of attack. As the recent piloted spaceflights by the Scaled Composites team have shown, modest investment (relative to that spent by major governments) can even put these altitudes within the grasp of non-government interests.

The major impediment to this type of attack is the requirement of getting the cloud of debris into the path of a satellite so that the convergence of the two trajectories results in a destructive impact. Though the fine degree of missile flight control shares much with ICBM development, the space tracking needed to locate a target satellite and calculate its motion with exacting detail is hard to obtain. This, of course, can be overcome with wider areas of attack (putting more debris in space), but as the low incidence of satellites being disabled by natural and artificial space debris demonstrates, the chances of such a primitive ASAT successfully hitting its target before its ballistic course brings it back to Earth are marginal.

Terminal guidance and manoeuvring, such as was used by Soviet and US ASAT experiments, or an extremely accurate space tracking system would be necessary for more effective attack by physical kill mechanisms. Depending on how close the interceptor has to be with the target before the chosen kill mechanism is lethal, the miniaturization of a nuclear warhead would perhaps be the better investment for the belligerent nation.

Scorched Earth style space denial is potentially inexpensive compared to other space power doctrines that involve weapons in space (those that involve some precision in their attack for instance). However the emergence of such a threat may be responded to along two primary schools of doctrine: survivability and space control (space force protection).

Space survivability is simply to make an attack against satellite infrastructure bearable by making satellites able to withstand attack, by having the ability to replenish satellites in orbit as quickly as they fall, and ultimately being able to do without the services offered by lost satellites.[18] Survivability emphasises that satellites are highly vulnerable to attack, perhaps more so than assets in other mediums.[19] This doctrine also presumes a continued lead in US space capabilities as in its purest form it does not address the deployment of US counter space forces.

Effective alternatives to satellites are envisioned by some as being a deterrent to space attack. Satellite alternatives, including theatre range Unmanned Aerial Vehicles (UAVs) acting as pseudo satellites,[20] which allow advanced US forces to continue operating effectively, would functionally counter or make irrelevant an opponent's counter space activities. As a space attack would not cause significant impediment to US operations, then according to this line of thought, there would be no reason for an adversary to invest in counter space capabilities. This is a calling to reduce the US dependence (and hence the vulnerability dependence creates) on space systems.

Earth bound systems, however, cannot completely supplant vulnerable space systems. Taking advantage of the vantage point of space is for some functions the only cost-effective means of their provision. Launch costs severely limit the amount of protection or hardening a satellite may have for withstanding an attack. Depending on the threats being faced, hardening and other forms of "armour" may even be the least cost effective solution. Countermeasures to attack must be tailored to the weapons being faced. Due to the numerous ways a satellite may be "attacked" a satellite may have to incorporate multiple types of hardening and other defensive systems. The cost of hardening these satellites must be compared to the cost of doing without a service, having redundancies, and ultimately finding ways to avoid being hit during an attack.

In certain terrestrial cases, avoidance of hostile fire has meant active interception of inbound munitions. Missile interception schemes have been scaled down to serve as a substitute for armour thickness in armoured vehicles.[21] As munitions in the traditional sense are not the only potential threats to satellites, a more layered approach is needed. Doctrinally, satellite defence plans could come to resemble naval task groups by surrounding critical service provision satellites with dedicated defence satellites when hardening against attack is no longer feasible.[22] This could potentially expand the defensive zone for a layered counter-counter space capability to be all of LEO space. This then leads to the costs and benefits of controlling the space environment via more active means.

Objectively, space warfare missions are analogous with those for sea, air and even land warfare environments. Even the most controversial of military roles currently given serious discussion (the space to surface attack) is analogous with conventional aerial bombing, artillery, or naval to shore strike missions. In addition, the ability to precisely tailor weapons effects allowing for either temporary disablement or vastly reduced collateral damage is permeating into space warfare thought. It is therefore no surprise that there exist arguments in favour of treating space as being another environment for warfare, another environment to be controlled as part of military operations.

Space control is rapidly emerging as the dominant strategy for the US in space, although it hasn't quite supplanted space as sanctuary in practice. Though many forms of satellite and space threat negation methods proposed as part of space control do not necessitate the clear use of space weapons,[23] this strategy is currently bringing the issue area of space weaponization to attention. As defined by the US Strategic Command, space control is:

> ...the ability to ensure our use of space while denying the use to our adversaries, if required. Space control consists of four functions: surveillance of space; protection of space forces from hostile threats and environmental hazards; prevention of unauthorized exploitation of U.S. space capabilities; and -- if required – negation of space systems hostile to the U.S. and its allies. [24]

Space control (or space superiority) is a military capability that will not likely be bestowed on the United States by negotiation or an international body. Presently, space capabilities are relatively safe by default, as few nations really have the ability to invest in abilities symmetrical to those the US possesses in space. For similar reasons, short of nuclear exchange, no nation has the ability to threaten US space capabilities on a wide scale. Unless there are massive changes to both the domestic and international situation, the preference will be to maintain this superiority.

The continued dominance of space by the US and the security it confers is under potential threat from its very success. Many major and minor powers have fielded military communications and reconnaissance satellites. Though facing many financial challenges, Russia is not short on satellite and spaceflight expertise, and continues to maintain a military presence in space.[25] China and Europe are also expanding their capabilities. US success in using space to enhance its military forces on the ground has certainly not gone unnoticed.

Even minor powers, seeking elevation of status, are pursuing space capabilities. Despite ongoing social and economic difficulties, Brazil, India, and Iran[26] are all pursuing space programs. Having a national spaceflight capability appears to be akin to the race among minor powers to build battleships that occurred in the era Alfred T. Mahan wrote his theories on sea power. Brazil was one of these 19th and early 20th century naval arms racers. Another medium power seeking ascendance during the Victorian era warship race was Japan, whose battle fleet was able to triumph over the great power of Tsarist Russia in the Battle of Tsushima (May 27, 1905). Japan has an active space program and is occasionally forwarded as a potential rival to US space superiority.[27]

Notwithstanding tensions with Iran and North Korea, nations with existing and emerging space capabilities are not presently engaged in military confrontation with the US. With the case of Europe and Japan, these are contemporary allies of the US. Despite the historical precedence of ally and adversary status being only temporary, recognizing this fact openly in policy is somewhat un-diplomatic. Certainty there is unease even in the war games used as tools to develop strategy and doctrine when real nations are implied as being future foes due to their peer, or near-peer, status.[28] There is also the issue of discussing in the open the possible need to neutralize the property of nominal allies that do not wish to get directly involved in US conflicts or that of neutrals.

Neutral players of course include private companies and potentially other non-governmental organizations. Lately, greater reliance has been placed on commercial assets by the US. Due to the 2003 US Commercial Remote Sensing Policy, government agencies, including military and intelligence, have been directed to make use of commercial services for their space to surface observation needs.[29] Not all of the companies and consortiums with militarily useful space capabilities have the US as their exclusive customer. Nor are they all under American jurisdiction. A possibility exists that such services may be provided to nations and organizations hostile to US intentions.

Controlling knowledge becomes more and more difficult with the proliferation of sources. As more nations deploy comparable space assets such as navigation and imaging satellites, the threat of having a peer or near peer competitor acquire RMA like capabilities grows. Even the space programs of US allies, such as Europe's proposed *Galileo* system, is a source of irritation and outright alarm for the US defence establishment since it offers a civilian GPS-like capability outside of their control.[30] Chinese investment in *Galileo*[31] only adds to fears that it will someday be used against US interests.

As the United States has no way to guarantee that space systems provided by vendors outside of its control will be used against its interests, the ability to unilaterally deny these services to others becomes attractive. In the future, control of space may have to be obtained through use, or threat, of coercive force. Proliferation of technology and the very success of US space operations thus far encourages others to seek either similar capabilities, or seek to deny the US an advantage in space. This fact is recognized by the top foundational doctrine statement of Air Force Doctrine Document (AFDD) 2.2.1, "US Air Force counterspace operations are the ways and means by which the Air Force achieves and maintains space superiority."[32] Unless the US gives up on space superiority as the background of operations, it is unlikely that it will negotiate away the means of obtaining it. This leads to the issue of space weaponization, and the problem of technological proliferation.

From a realist standpoint, an effective negotiated prohibition on interfering with one another's space capabilities would provide an opening for others to narrow the gap with the US in relative power terms. For a power that could afford it, space capabilities symmetrical to those proven by the US could be procured. By this, the US would lose the advantage that its heavy emphasis on space systems provides to its war fighting ability. There would be no lead if everyone held the same capabilities. As security is dependent of relative capabilities instead of absolute terms, the United States would be left more insecure. Of course this situation requires that US space capabilities cease to develop themselves while others catch up, though copying a capability would tend to be cheaper than pushing the envelope of what could be done.

On the other side, no other great power would willingly accept the hypothetical notion of an agreement that would have its potential space capabilities limited before parity in space systems capabilities is reached with the US. Though the US lead in space exploitation is relatively great, it is not unassailable as demonstrated by concerns over (non-weapon) military space efforts in China, Russia and Europe. The so-called revolutionary capabilities being demonstrated by US and allied forces are fast becoming mainstream military thought. Unless a competing power is content to go a different route (implying a "space free" RMA to compete with what is currently thought of as the RMA) its military leadership would be loath to give up on proven military force structures.

This is the problem: US security is presently based in large part on maintaining a significant lead in military power. Nations that are military competitors and perhaps challengers to the US would have their own security increased if they could replicate US RMA capabilities, thereby reducing the gap with the US (if the US's own power remained static). On its own merits, neither would accept a treaty that would allow the other to achieve capability gains relative to one another. That is not to say that changes in relative power cannot occur peacefully. The quiet collapse of the Soviet Union demonstrated that once funds became scarce it could easily accept a reduction in its power relative to its old Cold War rival. However, if given a choice (and it seems that funding is the key factor here), the US would not accept a ban of space weapons on terms that allow its competitors to potentially use space for the same military purposes.

Policy wise, the question of what to do about allied and neutral space systems that threaten US interests has already been answered. Many cite the August 2004 Air Force Doctrine Document (AFDD) 2-2.1, *Counterspace Operations*, which not only hints of a policy of weaponization, but also makes clear that third party space systems providing services to an adversary may be considered legitimate targets.[33] There is an element of sabre rattling here, as such statements serve as warning to third party space operators to avoid providing services to nations and groups the US considers threatening. While it does not elaborate on details, it does represent a change of the Eisenhower era sanctuary position, with the declared threat to interfere with foreign satellites that are deemed harmful.

This document also acknowledges the risks of taking such drastic steps: "...counterspace operations against adversaries using third party space capabilities may have economic, diplomatic, and political implications."[34] The emergence of "total war," has often meant that often no target can be spared if its value is great enough. Unrestricted submarine warfare in the two world wars demonstrates this concept at sea. It must also be remembered that total war has consequences as well, and is generally not the first option in dealing with a crisis. Certainly warfare against satellites must be carefully weighed against its many costs, not the least of which is political fallout. With respect to the decision to employ military force, few decisions are free of controversy.

The timing for active space control development is for many the true debate. Perceptions as to the extent that space technology has proliferated fuel a debate as to how far in the future active space control will become a clear and present requirement for the US military establishment. However, for some, the time to develop and deploy a robust active space control capability is far sooner. It is suggested by some that there is no requirement to forecast when foreign space capabilities will become a problem, as the purpose of US space weaponization will be to constrain and control what other nations may do in space.

Hegemony, High Ground, and Common Ground

Space is a realm over which no nation currently has claims of sovereignty, but it is of value to many. It is not certain that nations will indefinitely abstain from claims of sovereignty beyond the 100 kilometre altitude limit. Though it still invokes the "giggle factor" (or bemused alarm in some internationalist-multilateralist circles), there are the occasional urges to codify in legislation territorial claims on natural and manufactured celestial born habitats. Such was the case of the *Luna 2010* proposal for lunar settlement and eventual status as the fifty-first state in the United States.[35] Though this was a challenge for space development, it also highlights that cooperation is not the only option available for space policy. Indeed, with space being described as a "frontier," one must also remember all the wars fought between the great powers of Europe in the "frontier" of North and South America.

Common territory in terms of international relations, specifically with regards to Great Power strategy, is viewed with greater exclusivity. Here space policy and doctrine draws parallels with the writings of Alfred T. Mahan on command of the sea. Controlling the common territories is control over the common linkages that allow the modern global economy to function and for power to be projected.

For centuries, command and control of the sea by a state whose interests include relative safety for commercial shipping have allowed that state to become predominant. Along with generating the wealth needed for great power status, control of commons prevents it from being a vector for attack and invasion, and in turn allows its use as a medium for power projection. Though many, especially those not deriving the maximum benefits of naval hegemony will disagree, it can be claimed:

> Domination of the seas by a great maritime power in the cause of economic and thus political stability has resulted in protracted periods of seeming 'peace.' Each so-called *Pax – Romana, Britannica* and *Americanna* – has really been naval peace, where supremacy at sea provides a major deterrent against serious challenge by unfriendly opponents [Emphasis in original].[36]

Space is the ultimate common ground, and the ultimate high ground. Command of the Medium theorists such as Guilio Douhet for airpower, and Mahan for seapower, have suggested that from command of their respective medium of interest, all others could be controlled. A maritime nation that chose to exploit its favourable location and other characteristics uses the sea as a defence against invaders, as a conduit for military power projections, and for commerce. Railroads and later road networks were argued as offering similar capabilities for land powers. Bulk transport at sea however retains its critical role as a conduit for commercial and military aspects of national power.

Aviation, while not historically being such a great conduit for commerce as the oceans, quite clearly does act as a conduit for the projection of military power. Aircraft bring a third dimension to terrestrial movement, and by doing so may have constituted an RMA. This horizontal component makes airpower, unlike seapower, applicable across the entire surface of the Earth.[37] Elements of maritime power, warships and merchant vessels, have been for some time now under threat from airpower. Arguably, airpower has the capability to take on some of the roles of sea power but with generally lower cost platforms.[38]

Space, due to the already mentioned difficulties of getting and operating there is even less of a conduit for traditional bulk commerce. However, advocates of a more hegemonic space policy do make clear that space surrounds all other mediums of human activity. From space, mobile targets such as ships, aircrafts and missiles can be located and tracked. If a target can be seen, then it can be attacked. In an era of joint and networked warfare, targets can be handed down to terrestrial weapons, thus avoiding the significant engineering problems of a space to moving-terrestrial-target capability. Notwithstanding the problem of reducing the considerable costs, space power, like airpower before, promises the ability to again strike against the maritime foundation of the global economy:

> Whoever controls space, therefore, will control the world's oceans. Whoever controls the oceans will control the patterns of global commerce. Whoever controls the patterns of global commerce will be the wealthiest power in the world. Whoever is the wealthiest power in the world will be able to control space."[39]

Space, for many in the security and defence community, is the ultimate "high ground." The vantage point of space offers a commanding view of all that happens on the surface of the Earth. Advantages of possessing such a view have been well proven by the militarization of space up to and including its influence on the RMA process. However, beyond safeguarding RMA capabilities, space offers even greater potential. Ultimately, advances in space technology promise for some the ability to directly project power against any target within the Earth's atmosphere.

For these reasons, seizing and exploiting space is likened to the imperative to seize elevated positions in ground war. The "high-ground" label is derived from this focus on assets being able to project power down from space as if from a strategically important hilltop. Space with respect to potential great power competitions becomes under this type of logic, the only battlefield of consequence. All that can be detected, can be attacked, and there is not much that may hide from orbiting sensors.

This view of space power has also been labelled the "hegemonic"[40] view by Karl P. Mueller. A sufficient preponderance of power would go beyond being a deterrent and allow the hegemon the freedom to exercise its will and create a strategic stability in the shape of its own desires. Conceptually holding the high-ground would seem to be a natural part of maintaining a preponderance of power. Holding space for the purposes of dominating all earthly battlefields and indeed the international system would be hegemonic in nature.

There is a certain idealism found in the most aggressive vision of US space weaponization. The mainstream view among proponents of the US seizing and controlling space as rapidly as possible, is that the role of the United States in history has been largely benign (and even beneficial) to the world. Any move towards increasing its preponderance of power is seen as being not just for its own security, but more as an important tool in creating a "better world." Unquestioned space control becomes a "public good"[41] for the world for those that propose this type of space militarization.

Optimistically, if space is the critical battleground for the major powers, then seizing total control of space before any near-peer can challenge it would in itself deter all others from seeking space weapons. Controlling the space immediately surrounding Earth, Low Earth Orbit being the most stable vantage point, would be control over what was allowed into space and consequently prevent another nation from participating in a space arms race.[42] Space dominance would allow a benevolent United States to reshape a "better world."[43] Heavy investment in space weapons systems, especially after the end of the Cold War, in the words of US Senator Bob Smith, "would be a real 'peace dividend' – it would actually help keep the peace."[44]

Failure to immediately secure space militarily would at the very least also allow an arms race to occur as other nations attempt to counter US plans. At worse it is argued that space could be lost and along with it US predominance in the world. Proponents of High Ground doctrines accept that technology will eventually allow mass deployment of space weapons. The concern for space dominance advocates is the question of which nation will develop the technology, and consequently be first to seize the "high-ground." Without going into the intricacies of how elites in the US view the intentions of other national elites, one would suspect that those advocating hegemonic power would not want another power to have it instead of the US. Therefore, speeding along development and actual deployment of space weapons becomes paramount.

Realpolitik often spells the end of idealism, and that may be true here of the cited benefits to US security promised by the rapid achievement of a dominant command of space. Space weapons, even with increased funding, will take time to fully develop. Like the option of a massive deterrent capability, development of large scale space weaponization will be difficult to hide. Conspicuous aggressive space weaponization could encourage other nations to pursue space weapons. This in turn would give further reason for the US to accelerate its own space weapons program. In effect the seeking of security by attempting to seize space dominance, the need for space weapons becomes a self-fulfilling prophesy due to the security-dilemma it could cause among competitors. From such a situation, fears of a new and costly arms race emerge.

The US does not have a monopoly on space technology. Indeed with high profile difficulties experienced by NASA and others in the US aerospace community over the space shuttle and the long saga of attempting to replace it, doubts are beginning to emerge over the size of the lead enjoyed by the US government in spaceflight technology.

With respect to the reaction of US allies over a doctrine that emphasises dominant command and control of space, there are once again doubts over the notion that such a move would be appreciated by them. Given a great enough disparity in capabilities between allies, insecurities may develop in the weaker parties regarding the intentions of the stronger. Moves to increase security (by the hegemon) can also be interpreted as, "unilateralist America pursuing its own military advantage at the expense of other countries, most of which do not favour putting weapons in space."⁴⁵ Depending on whether one places faith in self-help or multilateralism, the weakening or strengthening of relations between allies affects security differently.

At the far end of possible allied behaviour, alliance forming or "balancing" behaviour among nations may result in allies of the United States turning on it to form a balance against perceived unilateralist intentions. To some extent this fear of allies interfering with US freedom of action is already manifested in the denial of military and political support for controversial policy decisions and operations. These fears have in turn become the rationale for the US desire for a global range strike capabilities. Basing strike weapons in neutral space would certainly overcome the problem of gaining over-flight and basing rights from foreign nations, often with their own agendas in a crisis.

Ballistic Missile Defence

Related to high-ground doctrine is BMD's involvement in space weaponization. Missile defence does not necessarily have to involve the placement of weapons in space, however due to its vantage point, orbital basing yields superior results for most BMD concepts. The vantage point of space is certainly useful for less provocative BMD systems meant for missile detection and tracking.

BMD is in some senses a strategic capability in that its existence is meant to have consequences on the policy options available to both the US and to its adversaries. Arguably by limiting the effectiveness of ballistic missile attack, a BMD system would compel nations to pursue weapons policies more acceptable to the US. By being a strategic capability, able to significantly affect if not dictate the (security) policy choices available to an adversary, there is indeed cause to label BMD a hegemonic or high-ground capability.

Until recently, most attention on space weapons related to the question of ballistic missile defence. The vantage point of space will always promise many inherent benefits (and costs) for the BMD mission. For these reasons, the space weapons debate is linked to the BMD debate. The Strategic Defense Initiative (SDI), arguably brought space weapons first to prominence as a policy debate. This debate was framed in a time when nuclear deterrence was the predominate concern of the US security and defence community. The debate of SDI was certainly framed by the strategic situation nuclear weapons produced.

Though most often thought of as an escape from the problem of nuclear deterrence, there are some who argue that SDI would have actually continued the viability of Mutually Assured Destruction (MAD) on which deterrence was based.⁴⁶ Missile defence, as most of its supporters and seemingly all of its critics contend, cannot stop enough warheads in a massive nuclear attack to prevent the damage that defines "assured destruction."[47] However, missile defence can, if credible, put into doubt that a first strike will be able to destroy retaliatory forces. During the Cold War, this would have meant that "that a Soviet war planner could not successfully plan for a first strike, since he would be unsure of how effective his first strike could be."[48] It is the knowledge that a massive nuclear exchange would be the expected result of nuclear attack, resulting in unacceptable costs to the initiator is the basis of nuclear deterrence.

An alternative systemic argument made in favour of SDI is that it made irrelevant the existing situation of deterrence based on MAD. Somewhat optimistically, SDI for some of its promoters became a way of escaping deterrence strategy. Deterrence based on the threat of immediate and total annihilation (as defined by MAD) as a way of great power interaction was morally troublesome for some in power. SDI's hope for escaping this dilemma's imposed by nuclear technology can be found in President Ronald Reagan's March 23, 1983 Address to the Nation on Defense and National Security:

> What if free people could live secure in the knowledge that their security did not rest upon the threat of instant U.S. retaliation to deter a Soviet attack, that we could intercept and destroy strategic ballistic missiles before they reached our own soil or that of our allies?[49]

SDI was meant to control the strategic environment of the Cold War. Idealistically some hoped it would allow the downplaying of nuclear weapons in the superpower relationship. Even if it could not break the US from a deterrence posture, it could at least prevent the Soviet Union from achieving a situation where a first strike posture would be potentially attractive. Missile defence, made possible by technology, would be used to prevent a detrimental security environment brought on by the technology of better ballistic missiles.

With the end of the Cold War, there are fears that nuclear and other Weapons of Mass Destruction will proliferate into the hands of nations that will not show the restraint of those that now possess nuclear weapons. Nuclear deterrence, though problematic for many world leaders, was the predominant strategy among the superpowers and other nuclear weapons states of the Cold War era. There is no guarantee that the same strategies are shared by those seeking to gain nuclear weapons. Furthermore, the possession of even limited deterrent capabilities by so-called "rouge states" hostile to the United States would limit its freedom to pursue objectives in some regions of the world. Along with the goal of guarding against accidental launch by one of the traditional nuclear powers, missile defence is envisioned to alleviate some of the fears that regional powers and (even less likely) subnational groups would be able to deter US operations with ballistic missiles.

A credible missile defence system would put in doubt the ability of a small ICBM arsenal to deliver even a single warhead. Such a defence could potentially minimize the impact of any nuclear challenge posed by a nation that has recently acquired such weapons. All that would be possibly accomplished after the great expense of acquiring a small strategic ballistic missile capability would be drawing criticism from the international community, without any significant decrease to US security.

A potential foe would be faced with two options, either it would have to acquire a deterrent force capable of the type of mass attack feared during the Cold War, or it would be better off not acquiring ballistic missiles armed with WMD.[50] The former would be prohibitively expensive, hard to conceal and may even cause further political costs, up to and including pre-emptive war to stop it from acquiring such capabilities once it became apparent it was seeking a large scale nuclear arsenal. The choice of pre-emptive war would be much easier if an ABM system was thought to be capable of handling the opponent's small nuclear arsenal.

In other analyse, certain types of BMD, boost phase defence in particular are related to space control. The space force protection component of space superiority would be vastly simpler if an adversary's space denial systems could be prevented from reaching space in the first place. Operationally there is no difference between an ICBM and a launch vehicle during the climb out of Earth's atmosphere, and therefore there is no difference in stopping a WMD payload or an ASAT payload. As mentioned before, the most common latent space weapons capability happens to be the nuclear armed ICBM that BMD is meant to counter. "Indeed, it is impossible to have a space control posture without having the capacity for national missile defense as well; the two capabilities are inextricably linked."[51]

All of the promise of missile defence ultimately rests on having an effective and credible system. While there is nothing in the realm of physics to stop the development of a credible ground-based boost phase interception system, a space-based system, as mentioned earlier in many respects would be superior. The 1993 USAF Air University study titled *Spacecast 2020* identified the reasons for investment in space power as being: Global View, Global Reach, and Global Power.[52] These inspiring words (for those that support a more hegemonic strategy) are easily applied as the requirements for missile defence system able to target threats emerging from any point on the globe.

Resurgence of Nuclear Weapons in the Tactical Role

Nuclear weapons are neither a particularly high technology as they have been around for about half a century, nor are they particularly accessible technology due to ongoing counter-proliferation efforts. This is the technology that some pundits have accused of threatening civilization. Coupled with the ICBM, these weapons are one of the primary threats against which BMD is meant to protect. Despite passing from the nuclear age to the present so-called information age, these weapons represent a particular threat to the information economy, and information warfare.

Any nation with missiles capable of reaching beyond the boundary of space and nuclear weapons technology can destroy spacecraft through the many immediate and long-term effects of a High Altitude Nuclear Explosion (HANE).[53] A nuclear kill mechanism is, however, not what the US presently has in mind when discussing missile defence or space weapons. The effects of such an attack would conflict with the notion of making space safe for US and allied operations. Space is crowded by many satellites and a nuclear event would have far reaching consequences. The accidental discovery of just how lethal and far ranging the effects of a HANE on electronics, whether earth-bound or in space, was found during a Project Starfish Prime nuclear test,[54] and has dampened enthusiasm for this type of counterspace option in the US. Due to the indiscriminate wide area effect of a nuclear detonation in space, this counterspace option is generally not discussed.

Perhaps it is better to say that presently in the US the nuclear option is not often discussed as a desired option for either missile defence or counter-space roles. For a nation hostile to the US, it has been suggested that the use of nuclear weapons against US (and all other) space infrastructure is an emerging threat. With the proliferation of missile/launch vehicle and nuclear weapons technology, this threat is not limited to near-peers of the United States such as Russia and China. Nations usually regarded as having only the ability to threaten regional neighbours, even with the acquisition of nuclear weapons, could threaten not just the US ability to intervene with modern (RMA type) forces, but also the modern information base economy.

> ...the potential that any nation acquiring nuclear-armed ballistic missiles will have not only a crude but effective space denial capability, but the power to hold the global economy at risk.[55]

The US Army's 1997 Winter War Game had the startling result of the RMA type force of 2020 (US expectations at the time for its military forces in 2020) being brought to the negotiating table by those simulating a near-peer that for various reasons (including protesting certain decisions made by the war game's judges) detonated several nuclear weapons in space.[56] This act of defiance effectively eliminated much of the globe's space infrastructure. The sudden loss of military space support effectively put an end to the RMA force (of 2020's) ability to respond militarily to its fictional opponent, and compelled it to seek negotiation. In light of some describing the purpose of this wargame as being a showcase for RMA concepts, it was something of a shock for the more advanced side being brought to the negotiating table.[57]

In addition to the problems caused to US defence planning, the losses to commercial space infrastructure due to a concerted HANE attack could possibly result in severe global economic upheaval beyond those caused by the US simply losing a military contest.[58] Aside from billions of dollars of hardware lost, the information economy would suddenly lose significant amounts of bandwidth, often to areas where satellite communications was the only cost-effective solution. It is uncertain (in that it has not happened before) as to what this kind of loss would mean for the economy. The often cited disruption of services from the Galaxy 4 communications satellite in 1998 from natural solar activity is estimated to have been valued in the millions.[59] This was the effect from the sudden loss of only one satellite; a HANE event could possibly result in significantly larger losses.

As mentioned earlier, warfare in space (with its only population residing primarily on the International Space Station) has the attraction of few direct casualties. It is theorized that a large enough high altitude nuclear explosion would through relatively long lasting enhancement and reshaping of the Van Allen Radiation belt cause gradual degrading and eventual destruction of all unshielded satellites passing through its effects.[60] Naturally occurring charged particles in space cause gradual damage to satellites. Satellite launch is expensive, and satellites not expecting to pass through the Van Allen belt (or facing nearby nuclear explosions) are minimally shielded to reduce costs. A dramatic increase in the density (or amount) of charged particles and other radiation hazards would greatly increase satellite operation costs through shortening satellite life-spans or necessitating hardening.[61]

It is uncertain whether deterrence will be effective at stopping such an attack from occurring in the first place. Despite the severe military, economic and resultant social repercussions expected from as little as a single HANE event, nuclear retaliation against the perpetrator is not often suggested. The question of a response to a massive disruption of the US ability to function economically and militarily but without significant loss of life has been framed by some in "ethical and moralistic" language.[62] As threatening nuclear retaliation against the mass destruction of populations is still troublesome for some, threatening retaliation to avenge the loss of a few billion dollars worth of hardware would seem even less popular. At the same time, such an attack can also be framed as being an attack against the Western way of life, but that still does not resolve the question about what to do about such a threat.

A historical analogy to space warfare would be NATO's threat to use tactical nuclear weapons against the massive numerical superiority of Soviet armoured forces during the Cold War. The massive destructive capability of "tactical" nuclear weapons was thought necessary to counter the waves of Soviet forces expected to cross into the West during a war between the Superpowers. In addition, there was ambiguity over whether strategic nuclear weapons would be employed to protect Western Europe from being overrun. To the present day the US (and other NATO nuclear powers) continues to avoid adopting a "no-first use" policy when it comes to nuclear weapons.

It was only later that precision guided conventional munitions offered alternative means to counter the Soviet numerical advantage. The force multiplication or enhancement effect of what is now referred to as RMA equipment and doctrines in theory allows a numerically smaller force to overcome a larger conventional "pre-RMA" force. Much of the technology needed for Western forces is dependent on satellite provided services. This provides one weak point for a contemporary adversary to counter US technological advantage.

To counter mass Soviet armour, the West risked nuclear escalation with doctrine that promoted the use of tactical nuclear weapons. Only later did low cost precision anti-tank weapons threaten the pre-eminence of armour on the battlefield through conventional means.[63] As a counter to present day US military superiority through so called "information superiority," asymmetrical (in that the effect of the attack are vastly different from the force being countered) means are recognized as being necessary to strike at the weakest links of the information cycle between elements on the battlefield. Adopting a space denial strategy, whether it be based on nuclear or non-nuclear neutralization of US space services, is only one such option. Certainly finding alternatives tailored to challenge US conventional superiority is only limited by imagination.

There is irony in that satellite dependent transformational warfare is now put most at risk by the very nuclear weapons that they were meant to supersede as the pinnacle of US strategy. Over the last few post-Cold War US administrations, and especially with the current Bush administration, there has been a steady de-emphasis on nuclear weapons in favour of precision weapons and information superiority.[64] As mentioned many of the concepts included in the RMA were pioneered in the Cold War as qualitative alternatives to the tactical nuclear weapons option to counter the quantity advantage held by Soviet armour. A foe in the developing world would arrive at the technology needed for nuclear-weapons-based space denial capability well before it can even think of advanced space weapons such as micro-satellites, long range lasers, and kinetic kill vehicles. Essentially this represents a return of the tactical nuclear weapon, an asymmetrical weapon to use against the contemporary superior conventional forces.

Keeping nuclear warheads from reaching space would certainly be the best solution to such a threat. Both extremes of the debate offer idealistic hopes: extreme views on space as sanctuary (more towards pacifism) cooperation and disarmament will make such threats unthinkable to the extreme space hegemonic position that command and control of orbital space will give the US (or alternate great power) total control over what is allowed into space from the Earth. In between such views as before are more practical solutions than those extremes that tend to invoke the "giggle factor" rather than meaningful discourse.

Launch vehicle technology is quite clearly a dual-use capability. ICBM's are routinely converted to space launch vehicles, and launch vehicle technology is coveted for its military applications. Not all nations that covet ballistic missiles will wait for the ability to field missiles with all the purely weaponization qualities of robust construction, rapid launch cycle, and high acceleration. Simply getting off the ground with a warhead would seem to be adequate as is demonstrated by the sheer number of Scud derived liquid fuelled ballistic missiles in operation by many in the developing world.[65] Such technology is also applicable to space launch vehicles.

Stopping the proliferation of space launch technology would be one useful means of space infrastructure protection. However with the growing demand for commercial space access, and the nationalistic qualities of having a space program, stopping proliferation may be difficult and controversial. Restricting nuclear technology has become a difficult endeavour in light of claims that it is being used for "peaceful purposes." Launch vehicle technology, while clearly having military applications, does not have the same stigma as nuclear technology.

Proponents of missile defence claim that current missile technology regimes have only slowed the spread of such capabilities.[66] If commercial space access becomes more of a routine industry (in that it loses its experimental quality), potential economic development geared towards tapping this high-technology industry will become another avenue for proliferation of dual-use technology. Indeed, as more developing nations wish to enter into the 21st century economy, denial of such technology may be seen as interference by current space powers. Criticism of such regimes to limit rocket technology may even occur despite valid arguments that as developing nations, these governments should have priorities other than competing in the launch vehicle market. At the same time, the nationalistic qualities of spaceflight, prestige, and peaceful demonstration of weaponizable technology arguably contribute to the core reason for a states existence, security.

Stopping the proliferation of nuclear weapons to be sure has benefits that extend beyond the risks these devices pose to space infrastructure. This however is not a universally held position. The recent demonstrations of nuclear arms by both India and Pakistan, Israel's nuclear ambiguity, revelations by South Africa's post-apartheid government of a discontinued nuclear capability, plus the continuing fears that Iran and North Korea (among others) harbour nuclear programs, clearly shows not all are satisfied by the nuclear status quo. Fears of terrorists somehow acquiring such weapons are another indication that the end of the Cold War has not ended the nuclear threat by any means. In the end there is only so much that weapons control regimes can accomplish against those that covet nuclear weapons, arguably one of the few working deterrents.[67]

This however does bring back nuclear deterrence as being a possible counter. While nothing (policy wise) is impossible in politics, it would be unlikely for a nation to intentionally destroy the US military technology advantage without a reason. (This does not discount the possibility that a high-altitude nuclear demonstration may unintentionally disrupt US space systems).[68] Being able to carry out a debilitating attack against US space systems would likely be part of a grander strategy on the Earth's surface. Like the Japanese attack on Pearl Harbour, the destruction of space infrastructure would likely be for the purposes of preventing US intervention somewhere on Earth. If it was thought necessary to risk escalation and consequences of a space attack, then surely whatever the adversary's actual goal is, it must be important to the US. At the same time however, such goals must be thought outside of those the US is willing to protect under its nuclear umbrella.

This of course brings up the question of what geopolitical goals would allow an adversary to get away with the mass destruction of US space capabilities, yet not invite a nuclear attack. As mentioned before, with the exception of mass investment in conventional space weapons that not even the US possesses, nuclear warheads would be the best space denial option to use against the US. Already the offending nation would have brought upon itself the concern and possibly wrath of the international community with the use of nuclear weapons (and in space on top of that supposed "taboo"). After this, it would of course have to achieve its goals on Earth. These goals could of course be threatened by the US strategic arsenal.

Certainly there is the risk of nuclear escalation, if the US was confronted by a near-peer, such as Russia or China (both with their own nuclear deterrence forces), with a nuclear based space denial attack. Loss of US space infrastructure would not impede present day US strategic nuclear forces. Indeed many systems were designed to function in the aftermath of global nuclear war. At the same time one would have to wonder about what goal a near-peer, short of claiming hegemony for itself, would cause it to risk escalation. The 1997 winter war game had certainly such a scenario. Through an artificially generated history up to 2020, RED team (Russia of 2020 in all but name) found itself in a situation where it was both useful to negate US space power decisively through mass use of HANE's, to achieve a goal outside the US nuclear umbrella, invasion of a non-NATO country in central Europe.[69] It is questionable as to how likely this scenario would occur in the real world, though the scenario described in the 1997 Winter War Game and opposition mentioned earlier to nuclear retaliation does bring back the old problem of using nuclear weapons to protect interests other than protecting the homeland (especially if such an attack puts at risk the homeland).

Contemporary arguments in favour of placing arms in space include the possibility that less than near-peers will see utility in space denial. It would be easier for a near-peer to field space weapons against the US, but conversely a near-peer would likely have more investment in space infrastructure (though not as much as the US). Sanctuary and arms control arguments would have as much if not more relevance to a near-peer fearing US abilities to counter its attempt to catch up in space force enhancement systems. Less than near peers, without heavy investment in space infrastructure might see no reason to allow the US the advantages of space.

Nuclear weapons would be the first effective space weapon a developing nation would have access to. Unlike mass debris attacks, a nuclear weapon does not have to rely on probability once it arrives at its place of detonation. There is no need for advanced sensors and computers for terminal guidance. Reusability is not a design factor, and the very effectiveness of a nuclear warhead reduces the need to for a second attack on a target satellite. A nuclear weapon set to go off at high altitude would be the ideal "poor" nations ASAT.

Nuclear technology and ballistic missile technology may put the US vulnerability to space denial within the grasp of some developing world adversary. However, those that fear a "space Pearl Harbour" may be overlooking the whole issue of under what circumstances does it make sense for a greatly lesser power to use space denial?

There are certainly other means than space denial to counter US technologically dependent forces. Insurgency and other forms of asymmetrical warfare (different from US force structures) are being used by present day foes in Afghanistan and Iraq. It remains to be seen how successful the US will be able to use technology, as opposed to other means, to counter this type of adversary. Afghanistan did not even have the capability to even threaten US Airpower, let alone space power. Indeed, the "giggle-factor" associated with space weapons would likely cause a hostile leader to overlook such an outlandish plan to take away the ability for the US to field its RMA force. Nuclear weapons and ballistic missiles are expensive, and are unlikely to be developed for a purely space denial role.

In the later example, Iraq had conventional forces, which in 2003 went up against a relatively small US force wielding many RMA type systems including satellite-guided munitions, unprecedented amounts of satellite communications bandwidth,[70] and GPS navigation in the desert. Iraq's conventional force lost the fight. Now elements from both inside and outside of Iraq have continued to harass US operations through low-cost insurgent warfare. If successful, these low-cost (these are non-state actors now as the primary foe) asymmetrical means employed by insurgents may be more of a template for countering RMA style forces than expensive space warfare capabilities.

On the other hand, if the regime of Saddam Hussein had possessed the nuclear weapons and ballistic missiles it was accused of attempting to develop, perhaps things would have happened differently. The combination of nuclear weapons and an appreciation of what an attack against US satellites could do to US conventional war fighting probably would not helped the regime in the long run (especially since such an act would solidify international support against it), but would have helped to even the odds for its conventional forces.

This brings up a significant problem with many weapons systems: there is no guarantee that after significant investment to build up a capability that it will provide the right kind of ability to handle future threats. Long development times for advanced systems mean that for a weapon to be able to meet a threat its development must start years if not decades in advance. In the meantime, requirements, allies and adversaries may have changed. A weapon does not actually have to be used militarily to serve a political purpose. No nuclear weapons have been used in anger since the end of the Second World War, yet these systems still serve an important (though reduced) role in US strategy. If anything the existence of a credible missile defence system could be seen as a deterrent to developing nations spending their limited funds on ballistic missile development. While it is better not to have to worry about defending space infrastructure because of a lack of threats, the very success of the RMA style warfare it allows may invite challenges to its future safety. Ultimately weapons systems in space may have to be the last line of defence when diplomacy and reason fail.

For the question of space weapons deployment, it then becomes a question of what types of capabilities would cover the greatest range of threats, and reduce identified vulnerabilities. Near term vulnerability in space is largely confined to latent counter space abilities in nuclear weapons on long range ballistic missiles. Though the possibility for near-term exploitation of the US space infrastructure's vulnerability is remote, what will happen in future is hard to discern. Force protection as a potential first step to space weaponization has the benefit of being able to fit in with the long standing (but relatively unfulfilled) requirement for missile defence.

Specifically of importance to the US's desire to protect its forces in space is the development of a boost-phase anti-missile system. The arcing path of ballistic missile flight is divided generally into three phases. Boost-phase is the initial portion of flight where the missile accelerates to achieve the kinetic energy to carry the warhead(s) through midcourse; and finally, terminal or re-entry phase of its flight. A missile interception during early boost-phase would prevent a weapon from gaining enough momentum to carry it into space where a nuclear warhead could cause damage to nearby satellites (and potentially the theorized lingering after effects on all other satellites in Low Earth Orbit).

Aside from their defensive nature, both space force protection and ballistic missile defence missions may lead to the same equipment. Missile defence and space force protection would be one in the same when used to prevent nuclear weapons from reaching space. Indeed if one were to pursue total command of space, a boost phase capability could be used to prevent any unwanted spacecraft from reaching orbit. This however would only raise the anxieties of the international community.

Notes

1 Matthew Mowthorpe, *The Militarization and Weaponization of Space*, 30-31.
2 Paul B Stares, *Space Weapons & US Strategy: Origins & Development*. (London: Croom Helm, 1985), 237.
3 Ibid.
4 With the end of the Cold War, the overhead reconnaissance needed for verification of strategic arms reductions between Russia and the United States may now be done by lower cost aircraft. This has only been possible due to the realization of Eisenhower's idea of an Open Skies Treaty, which came into force January 1st, 2002.
5 Bruce Berkowitz, *The New Face of War*, (Toronto: The Free Press, 2003), 154.
6 Federation of American Scientists. "FAS Calls for Alternatives to Weapons in Space," 8 October 2004. <http://fas.org/nuke/control/os/> (2004).
7 Michael E. O'Hanlon, *Neither Star Wars Nor Sanctuary*, 116.
8 Matthew Mowthorpe, *The Militarization and Weaponization of Space*, 197.
9 Ibid, 104.
10 International Campaign to Ban Landmines, "Treaty Members," 21 July 204. <http://www.icbl.org/treaty/members?eZSESSIDicbl=5076874f9a019e800212f27d01202eba> (2004).
11 George Friedman, and Meredith Friedman, *The Future of War*, 296.
12 Roger G. DeKok and Bob Preston, "Acquisition of Space Power for the New Millennium," in Peter L. Hays, James M. Smith, Alan R. Van Tassel, and Guy M. Walsh, eds., *Spacepower for a New Millennium* (New York: McGraw-Hill, 2000), 85.
13 Bernard Cornwell, *Historical Note*, in *Sharpe's Escape*. (London: HarperCollins, 2004), 347.
14 Jonathan S. Lockwood, "Space Control Versus Space Denial in 21st Century Warfare: Achilles Heel of the RMA (Revolution in Military Affairs)?" *Defense & Foreign Affairs Strategic Policy* 28, no.8 (2000):4-6.
15 The Commission to Assess United States National Security Space Management and Organization, *Report of The Commission to Assess United States National*

Security Space Management and Organization, 11 January 2001. <http://www.space.gov/docs/fullreport.pdf> (2004).

16. Daniel G. Dupont, "Nuclear Explosions in Orbit," *Scientific American* 290 no. 6 (July 2004): 107.
17. George Friedman, and Meredith Friedman, *The Future of War.* 364.
18. Michael E. O'Hanlon, *Neither Star Wars Nor Sanctuary,* 123
19. Lt. Col. (Retired), USAF, David E. Lupton, *On Space Warfare,* (Maxwell Air Force Base, Alabama: Air University Press, 1998). <http://www.airpower.maxwell.af.mil/airchronicles/apj/apj98/win98/deblois.html> (2004).
20. The pseudo satellite must of course be completely independent of space systems themselves. Many of the longer range UAV's rely on satellite communications themselves when the curvature of the Earth makes line-of-sight (LOS) transmission impossible.
21. George Friedman, and Meredith Friedman. *The Future of War.* (New York: St. Martin's Griffin, 1996), 154.
22. It is of note that warships have largely abandoned armour in the face of contemporary threats. With missiles being a greater threat than large calibre shells, anti-air defences have arguably shown more utility than armour plate. Despite modern developments, proponents of the armoured battleship still linger.
23. As mentioned before, electronic warfare techniques that do not leave lasting effects (signals jamming) are not generally considered weapons.
24. United States Strategic Command. *Space Missions,* March 2004. <http://www.stratcom.mil/factsheetshtml/spacemissions.htm> (2004).
25. Matthew Mowthorpe, *The Militarization and Weaponization of Space,* 76-77.
26. British Broadcasting Corporation, "Tehran aims for satellite launch," 5 January 2004. <http://news.bbc.co.uk/1/hi/world/middle_east/3370143.stm> (2004).
27. Everett C. Dolman, *Astropolitik.* 114
28. Jonathan S. Lockwood, "Space Control Versus Space Denial in 21st Century Warfare: Achilles Heel of the RMA (Revolution in Military Affairs)?" *Defense & Foreign Affairs Strategic Policy* 28, no.8 (2000):4.
29. Michael E. O'Hanlon, *Neither Star Wars Nor Sanctuary,* (Washington, DC: Brookings, Institution Press, 2004), 38.
30. Oliver Morton, "Europe's New Air War," *Wired Magazine,* August 2002. <http://wired.com./wired/archive/10.08/airwar.html> (2004).
31. British Broadcasting Corporation, "China joins EU's satellite network," 19 September 2003. <http://news.bbc.co.uk/2/hi/science/nature/3416231.stm> (2004).
32. United States Air Force. *Air Force Doctrine Document* 2-2.1, 2 August 2004. <http://www.dtic.mil/doctrine/jel/service_pubs/afdd2_2_1.pdf> (2004).
33. Theresa Hitchens, "Europe's USAF Counterspace Operation Doctrine: Questions Answered, Questioned Raised," *Center for Defense Information,* 4 October 2004. <http://www.cdi.org/program/document.cfm?DocumentID=2504&from_page=../index.cfm> (2004).
34. United States Air Force. *Air Force Doctrine Document* 2-2.1, 2 August 2004. <http://www.dtic.mil/doctrine/jel/service_pubs/afdd2_2_1.pdf> (2004).

35 Klerkx, *Lost in Space: The Fall of NASA and The Dream of a New Space Age*, 251.

36 Clark G. Reynolds, *Command of the Sea*, (New York: William Morrow & Company, Inc., 1974), 4.

37 Seapower advocates may of course counter by citing the need for aircraft carriers to bring airpower across the globe. Arguably there is a bit of inter-service rivalry in the USAF's desire for a global range bomber.

38 Major, USAF, Roy Walker and Captain, USAF, Larry Ridolfi, "Airpower's Role in Maritime Operations," *Air & Space Power Chronicles*. <http://www.airpower.maxwell.af.mil/airchronicles/cc/ridolfi.html> (2004).

39 George Friedman, and Meredith Friedman. *The Future of War*, 411.

40 Karl P Mueller, "Totem and Taboo: Depolarizing the Space Weaponization Debate."

41 Everett C. Dolman, *Astropolitik*, 159.

42 Ibid, 157.

43 Ibid.

44 Bob Smith, "The Challenge of Space Power," *Aerospace Power Journal*, Spring 1999. <http://www.airpower.maxwell.af.mil/airchronicles/apj/apj99/spr99/smith.html> (2004).

45 Michael E. O'Hanlon, *Neither Star Wars Nor Sanctuary*, (Washington, DC: Brookings, Institution Press, 2004), 16.

46 Matthew Mowthorpe, *The Militarization and Weaponization of Space*, 19.

47 Nuclear deterrence as a stabling influence is based on that the idea that no side believes itself able to emerge from a nuclear exchange without incurring unacceptable costs. The development of an effective first strike capability, (the ability to initiate and get away with a massive nuclear strike) would be destabilizing.

48 Matthew Mowthorpe, *The Militarization and Weaponization of Space*, 19.

49 Ronald Reagan, President, Speech, "Address to the Nation on Defense and National Security." (23 March 1983). <http://www.reagan.utexas.edu/resource/speeches/1983/32383d.htm> (2004).

50 Ideally for the United States a foe would pursue the latter option including the total abandonment of WMD technology. The choice to pursuing WMD with non-ballistic missile means of delivery is another question.

51 Jonathan S. Lockwood, "Space Control Versus Space Denial in 21st Century Warfare: Achilles Heel of the RMA (Revolution in Military Affairs)?" *Defense & Foreign Affairs Strategic Policy* 28, no.8 (2000):4-6.

52 United States Air Force Air University. *Spacecast 2020*, 23 February 1998. <http://www.au.af.mil/Spacecast/Spacecast.html> 2004.

53 The damage radius from the immediate heat and radiation effects of a nuclear detonation are measured in dozens if not hundreds of kilometres. (In space without atmosphere to transfer blast effects, physical damage is significantly reduced.) The electronics damaging electromagnetic pulse (EMP) generated by a nuclear event is another well known immediate effect. Other electromagnetic effects do not require line-of-site and can affect electronics across the horizon

from the HANE by reflecting off the Earth and atmospheric layers. Then there is the enhancement and reshaping of the Van Allen Belts caused by the "pumping" of damaging charged particles. See: Daniel G. Dupont, "Nuclear Explosions in Orbit." Scientific American 290 no. 6 (July 2004): 100-107.

54 Benjamin S. Lambeth, *Mastering the Ultimate High Ground: Next Steps in the Military Uses of Space.* RAND. <http://www.rand.org/publications/MR/MR1649/> 2004.

55 Jonathan S. Lockwood, "Space Control Versus Space Denial in 21st Century Warfare: Achilles Heel of the RMA (Revolution in Military Affairs)?" *Defense & Foreign Affairs Strategic Policy* 28, no.8 (2000):4-6.

56 Ibid.

57 Ibid.

58 Ibid.

59 Ibid.

60 Daniel G. Dupont, "Nuclear Explosions in Orbit," Scientific American 290 no. 6 (July 2004): 103.

61 The Commission to Assess United States National Security Space Management and Organization, *Report of The Commission to Assess United States National Security Space Management and Organization,* 11 January 2001; available at <http://www.space.gov/docs/fullreport.pdf> (2004).

62 Daniel G. Dupont, "Nuclear Explosions in Orbit," 107.

63 George Friedman, and Meredith Friedman, *The Future of War.* 158.

64 Department of Defense, *Findings of the Nuclear Posture Review.* 9 January 2002. <http://www.defenselink.mil/news/Jan2002/020109-D-6570C-001.pdf> 2004.

65 Global Security. "Nodong-1," <http://www.globalsecurity.org/wmd/world/dprk/nd-1.htm> (2004).

66 Current rational for ballistic missile defence, includes the threat from emerging and potential nuclear armed states.

67 Disproving the effectiveness of nuclear deterrence through a nuclear war, arguably would be somewhat less appealing than letting additional nations gaining a deterrence capability.

68 Daniel G. Dupont, "Nuclear Explosions in Orbit," Scientific American 290 no. 6 (July 2004): 107.

69 Jonathan S. Lockwood, "Space Control Versus Space Denial in 21st Century Warfare: Achilles Heel of the RMA (Revolution in Military Affairs)?" *Defense & Foreign Affairs Strategic Policy* 28, no.8 (2000):4-6.

70 Michael E. O'Hanlon, *Neither Star Wars Nor Sanctuary,* 4.

Conclusion

Space is for many a realm open to many possibilities. Policy is only one influence that shapes reality from possibility. The challenge for US policy makers is to facilitate a balance of activities in this realm that remains favourable to Western security and prosperity. Almost certainly, not all activities in space will be in the best interests of the US. Some problematic space utilization may be tolerable, subject to friendly debate, and others may be unquestionably clear and present dangers. Technology opens both opportunities and vulnerabilities, which are interchangeable depending on which side one finds oneself. As investment in space technology becomes increasingly important to contemporary Western society (including its ways of war), the conditions and pressures encouraging warfare in space grow as well. Possibilities do not necessarily translate into policy, but only serve as the backdrop for it.

Presently the US has space superiority by virtue of its unequalled access to space, and development of space force enhancement systems. Few of the United State's strategic competitors, let alone actual military foes, have the ability to exploit space as it does. The current so-called "Revolution in Military Affairs" is an expression of US military space power. Without space force enhancement, at present the most aggressive use of space by the military, the shape of cutting edge military power would be dramatically different. US military and economic powers are linked to space power, this even without weapons in space. This is space superiority, but only by default. At present few others can harness the resources needed to challenge the US in space along military, economic or scientific dimensions, let alone all critical elements of space power.

The dominance of US space power is however not a constant in the international system. Great power status is something which must be built and maintained, often at great cost. Space flight, though technically challenging, is within the grasp of practically all industrialized nations. There is a tendency to associate having an active space presence with global power and influence. Near-peers of the United States have to varying degrees their own indigenous space launch and space utilization capabilities. Other nations find it more useful to contract their access and space services from 3rd party providers. The "Highest of high grounds," has many stakeholders. Not all are satisfied with the distribution of capabilities orbiting the Earth.

Proliferation of technology has the (arguably unintended), consequence of lowering the bar required for nations to be able to threaten Western space interests. Near-peers are without exception stakeholders in space, and perhaps even share a common interest in keeping space free of hostilities. Not all nations that can potentially bring weapons to bear on orbiting infrastructure have heavy investment in space. Indeed, this very asymmetry, between the US and a non-technological power, would imply that such an opponent would have even more reason to deny space to the US than a near-peer. A non-space power would literally have "nothing to lose" in space if it initiated a space denial strategy.

Not all international disputes can be solved without bloodshed. Technology is potentially expanding the number of opponents who can feasibly disrupt the space superiority the US enjoys. The consequences of such hostile action in space for innocents and allied forces on the ground are much graver than the loss of a few expensive satellites. As long as military force is the ultimate guarantee that space will not provide aid to a foe, rationale for taking the militarization of space across the weaponization threshold will exist. In space, as on the ground, the utility of coercive force persists to the present day, and for the foreseeable future. Technology and politics will produce the capabilities and motivations for one nation to take conflict into space. This reality applies equally to the US, its allies, and its competitors. Coercive force is not the only tool used to produce security, though sometimes it is the only one worthwhile.

At the same time the international system is not based on a blind use of aggression. The question of space weaponization brings up age old problems of security dilemmas, arms control, and fundamental questions over how legitimate security concerns are perceived. A weapon becomes dysfunctional with its purpose of providing security if the opposition's response to the weapon's existence outweighs the security provided. In this regard, the arguments for and against weapons in space have as much to do with perceptions than with the mechanics of spaceflight. Perceptions on what goals are important, the means of achieving them, and ultimately the shape of the world in which they must be pursued shape all debates over security and defence.

Just as there is no one standard as to what constitutes a legitimate security concern, there is disagreement over what defines a space weapon. Without a hostile shot having ever being fired in space, questions arise as to how "peaceful" contemporary space activities are. Space force enhancement tactics have satellites participating in the attack of targets within the Earth's atmosphere. Arguments can be made that any one or combination of satellite, atmosphere bound munitions, or various Earth based schemes to counter US style space force enhancement earn the title of space weapon. The line between space militarization and space weaponization is an imprecise or "fuzzy" distinction. Practical space security policy leaves little room for all encompassing dismissal or acceptance of space weaponization.

For some the debate over space weaponization is a debate over whether space as a domain of human activity is any different from those found on Earth. Notwithstanding latent space warfare capabilities, the occasional space warfare experiment, and disagreement over what counts as a space weapon, the consensus seems to be that space is at present a "weapons-free" environment. This brings hope to some that space is perhaps "special," a place too important for mere political entities to fight in. Others equate space as offering a set of military advantages and vulnerabilities which can only lead to a worrying standoff arguably similar to the nuclear confrontation which plagued the world during the Cold War. On the other hand humanity's access to space has been relatively short. Any such "taboos" against space warfare may perhaps simply be a matter of technology and threats not yet converging. Space is clearly important to warfare on Earth, which for many implies that it is important enough to arm.

The Future, including the long term shape of space militarization is unknown. While politics is described as the "art of the possible," not all possibilities are brought into reality. Jules Verne and other visionaries of the Victorian era foresaw many of the possibilities airpower would have on the world. Only some of these possibilities have become realities. Even fewer of these airpower possibilities have become practical. Historical circumstance and technology have grounded such ideas as mighty lighter-than-air craft ruling global commerce and warfare. Those speculating on the future development of space power face similar prospects. Space power moving from the periphery of the security agenda to the mainstream will as it did for early airpower policy analysts only raise the stakes. Confronting the unknowns of space weapons and warfare will be necessary as space flight technology and the militarization cannot be "undiscovered."

APPENDIX ONE

Some Enabling Technologies

The placement of weapons into outer space is largely dependent on technology. Political will seems to be abundant in some quarters, but if the right mix of costs and capabilities cannot be engineered then the entire discussion remains in the realm of speculation and science fiction. Today it is possible to put things into space. It is possible to project destructive energies via lasers, kinetic energy, and other means. It is possible to have computers plot interceptions on objects traveling at orbital speeds. All of these and many other prerequisites for weaponization have been demonstrated successfully under "experimental" conditions. The only thing left is to integrate all these capabilities at a price which is appropriate for the threats expected.

The bleeding edge technology sparks the imagination at what can be done but is usually too expensive to put into practice. Deployment will require that the technology mature to the point of being robust and reasonably priced. While prognosticating the future of technology has an arguably less stellar track record than predicting international affairs, a few enabling technologies are identifiable. Though this is not a complete list, key technologies include: the shrinking computer (accompanied by ever increasing capability), low cost space access, and follow-on micro-satellites to the ones already causing controversy today. These three technologies promise to reduce the financial burdens associated with space weapons deployment.

Computers

There is perhaps no more critical a technology to any future space weapons prospects than the continued shrinking of microprocessors as they become more and more capable. Already much has been said about the importance of computers to modernity and of its surprising rise to dominance. At the dawn of the Space Age, early digital computers supported the various space programs, though human ingenuity, skill, and bravery held more prominence. In the Information Age, space technology only plays a supporting role along side fibre optics and other communications technology. Robotic space exploration is regularly suggested as a practical alternative to manned spaceflight. In this regard, computer technology and space technology have switched roles. With respect to warfare, it is only through the existence of low-cost micro electronics that precision attack is possible. As the type of space warfare capability being promoted within the Western defence community is one of flexible and precise effects, having available sufficient computer power would seem to be critical.

As a theatre of operations, space is presently unique in that the vast majority of military space programs do not actually put personnel into theatre. Satellites are mostly remotely controlled from ground stations. Few are capable of independent operation, let alone being able to recognize and respond to an attack. This need for ground control creates a large and expensive infrastructure on the ground, which itself may be targeted as part of potential space warfare strategies.[1] Dispersing or in the terminology of management, "empowering" assets to be able to take care of themselves would be a potential way of decreasing costs for military space systems. In addition, reduced reliance on human oversight for day-to-day management would reduce the chance that military satellites could be collectively endangered by attack on the handful of earthbound command and control facilities that maintain them.

The development of reliable autonomous satellite control is largely contingent on computer technology. NASA's *Deep Space 1* spacecraft's Remote Agent Experiment tested of this type of computer control in 1999.[2] With some proposed space weapons programs envisioning the management of hundreds if not thousands of armed satellites[3] in orbit at a time, there is a clear need to streamline the whole system. Beyond the navigation of a solitary deep space probe is the capability to maintain a large orbiting constellation, including perhaps the automatic movement of satellites to restore coverage in the aftermath of an attack. Computer systems that can orchestrate control over a large number of assets are being investigated not only for space applications, but also for the control of terrestrial warfare systems. Autonomous control over one asset remains difficult, let alone the coordination of hundreds, but the promise of this emerging technology continues to draw support and funding.

There are several benefits to having vast numbers of small (micro) satellites in orbit: smaller payloads do tend to be easier to launch, numbers imply reduced vulnerability, and by leveraging off today's computer technology each one of these satellites represents a group of resources that can be networked together to perform tasks that would require excessively larger spacecraft. From the same building blocks of cheaply available computer processors, high speed networks, and a flexible open architecture, cluster or parallel computers have demonstrated processing powers to rival more expensive supercomputers.[4] Of important note to military users is the ability of some cluster technology to allow the addition and subtraction (or failure) of processors without massive disruption to the work being done. Hundreds of orbiting Brilliant Pebble type interceptors represents a large amount of potential computer processing power, it is only a matter of designing into each satellite the capability to network together and share processing loads.

Operational issues such as distance between satellites, and communication lags may limit the scope of cooperation however even a network of all satellites within a small area may be enough to handle the processing loads associated space weapons applications.

Large amounts of satellites also represent potentially a large number of independent sensors which, through computer processing, can be combined to form a clearer picture of the battlefield than satellites several times larger. This is similar in concept to Synthetic Aperture Radars (SAR), where though the combination of a small moving radar and computer processing it is possible to "simulate" a much larger system.[5] There are proposals for orbiting radars that can provide all weather imaging reconnaissance, including those that have radars mounted on small micro-satellites.[6] Though the use of computers these independent radar images can be combined for greater clarity. Similarly, the thermal imaging sensors of hundreds of Brilliant Pebble type interceptors continually staring at Earth could be combined together to form a global surveillance system[7] in addition to being a missile defence system.

One possible use for all the potential computer power found in aggregating the spare computer power of a whole military satellite constellation would be the contentious issue of command and control of space weapons. Among the many criticisms over the various missile defence schemes proposed are the daunting computer requirements for target identification and discrimination. One of the principle attractions of boost-phase missile defence is that the target missile in boost-phase is a single large and easy to detect object. During the midcourse part of a ballistic missile attack, the warhead or multiple warheads separate from the missile in addition to potential decoys and other debris, all of which must be sorted out to allow for successful interception. Other computer needs include actually recognizing that a missile attack is underway, and plotting an interception whether it is by physical interceptors, or speed of light direct energy weapons. All of this must be accomplished quickly and reliably. Even the benefits of boost phase interception are under scrutiny in that boost phase while presenting an easier target is a much shorter portion of ICBM flight, limiting the chances of practical interception.[8] The need for quick reaction for missile defence in general has raised concerns over the inability for humans in the command chain to make these decisions, and the implications of allowing computers to do so on their own. Aside from concerns over the reliability of complex software, there is the problem of having enough processing power to actually run any software entrusted with protecting cities.

The mechanics of orbiting also present a similar need for effective computer command and control of weapons in the space control mission. Protecting against a surface to space ASAT would be similar to ballistic missile interception but perhaps with less time to react as the attack is taking place at the apogee (roughly midpoint) of a ballistic trajectory. Orbiting ASATs would present more time, but even this is limited as LEO satellites circle the Earth in minutes. An attack by a physical interceptor[9] against a satellite in LEO may require less than one full orbit as Soviet era co-orbital ASAT tests demonstrated.[10] The limits of human command and control raises the possibly that only computer control would be feasible for successful defence of friendly space assets. However before any computer system may be trusted with the defence of critical interests in space great strides in computer technology must be made.

Space Launch

Space weaponization implies having targets in and/or attacking from space. Therefore making space more assessable would be a primary factor into how far space becomes weaponized. At present the limiting factor to practically all space endeavours has been funding. Reaching orbital altitudes and velocities is a large part of this great cost. Any lowering of space launch costs will have the effect of making space weaponization more palatable.

Reduction of launch costs may also be accompanied by proliferation of launch technology, which arguably will cause an increased level of perceived threat to US Command of Space. While launch costs remain prohibitively high, the US as the last superpower has the option of outspending all challenges to its control of space in the near term. While a reduction in launch costs may be beneficial to American military use of space, any accompanying proliferation will increase the capabilities of competitors to replicate current space supplied military services and for them to potentially endanger US space infrastructure. With more access to space, there will be for some policy makers be increased need for more aggressive space control measures, including the potential for widespread deployment of purpose built space weapon systems.

Orbital access is primarily a matter of releasing huge amounts of energy in a precisely controlled manner. First, a launch vehicle must be able to generate a force large enough to overcome gravity and accelerate from rest. Reaction or thrust based propulsion systems (such as jets and rockets) are for the foreseeable future the primary means for getting a launch vehicle off the ground (notwithstanding the use of lighter-than-air "stages"). Newton's Third Law of Motion has that for every action there is an equal and opposite reaction. The thrust or force needed to overcome gravity and accelerate a rocket to some desired velocity is the product of expelling mass (propellant) in the direction opposite to travel:

$$\text{Force} = \text{Mass} \times \text{Acceleration}$$

For a given mass of propellant, the faster it can be expelled by an engine, the greater the reaction or thrust produced in the opposite direction. Newton's Third Law is only a beginning for all the concepts and math involved. Vehicle design, fuels, the nature of engine operations and other necessary technologies cover a broad range of science and engineering disciplines. A comprehensive examination of all that is involved is beyond the scope of this work.

Once off the ground, the orbital launch vehicle still must accelerate itself, payload and remaining onboard propellant to orbital velocities. Orbital velocity is dependent on altitude; lower orbits circle the Earth faster than higher ones though direct accent to high orbital altitudes is prohibitively expensive. Starting at some of the higher Medium Earth Orbits (MEO), transfer orbit techniques from LEO are the most fuel efficient method of orbit placement with current technology.[11] The direction of flight is also a factor in that the Earth's rotation imparts a bit of speed to the launch, though this is only true if the direction orbit actually has a major component along that of Earth's rotation. Equatorial launch sites are valued for this bonus push provided by the Earth. Direct accent to Polar Orbits makes little use of this extra push as this orbit is more or less perpendicular with the rotation of the Earth. This bonus becomes a penalty when directly ascending to an orbit that has a spacecraft travelling opposite to the Earth's spin. Finally there are safety concerns with over flight of populated areas, any launch vehicle or components shed during launch that do not reach orbital speeds will fall back to Earth on a ballistic flight path.

The present cost of space launch is often measured in units of thousands of dollars per pound (or kilogram). A large fraction of this cost is in the huge ground crew needed to prepare a launch vehicle for flight. Then there is the expense of constructing a vehicle which can generate and survive the tremendous forces associated with space launch, while being as light as possible. Finally there are the energy requirements themselves, a large amount of propellant is needed as a rocket based launch vehicle suffers from the fact that from its launch pad, the initial firing must generate enough energy to lift the payload, the launch vehicle and the fuel needed for continued acceleration. Once in flight, it becomes easier to maintain acceleration as fuel is burned lightening the whole vehicle. Dropping spent stages, boosters and other equipment needed only during lower flight regimes further lightens the mass being propelled to orbit. There is a balance between payload, vehicle and fuel which represents tradeoffs made in light of available technology and costs.

For various reasons, existing heavy launch vehicles do not follow well the engineering principle of keeping things simple. Describing the US Space Shuttle as the most complex flying machine in existence was once a point of pride for its backers. Now it is a criticism as that very complexity proving to be the shuttle's downfall. Many thousands of parts must operate correctly, and to assist with that near perfection a huge ground crew is needed to prepare and monitor these components from launch pad to orbit and back to the Earth. Despite this immense effort, accidents have already claimed two Space Shuttles and their crews. Other launch vehicles have fared no better, most competing expendable launch vehicles have suffered at least one major launch failure.[12] Space launch is a complex and dangerous technology, though reduction in complexity may improve reliability and safety and thereby reduce costs in the process.

The development of reliable computer diagnostics and flight control offers one potential avenue for reducing costs. The shuttle and most in use launch vehicles date from before the personal computer revolution. Since then computer assisted diagnostics have made their way into applications such as high performance warplanes (where cost reductions have been realized) and into automotives (where the effect on cost of ownership is arguable). Another use of computers is in launch vehicle flight control. Among the X-prize competitors, Armadillo Aerospace used rapid computer controlled thrust adjustments across its four engines to dispense with heavy mechanical means of keeping their vehicle pointed up.[13] Complex heavy mechanical systems and computer controls were replaced with just complex computers. In this case reliability and weight reduction were dependent of successful integration of sensors, fuel flow controls and computer software.

The actual propellants in common use do not have to be terribly exotic. Liquid hydrogen and liquid oxygen are commonly used in US and other rockets. Avoiding the problems of cryogenic propellant storage, less efficient oxidizers such as nitrous oxide (laughing gas) or high test hydrogen peroxide can be used in combination with a fuel like kerosene or other hydrocarbons. Solid fuels are often just aluminium particle fuel and a solid oxidizer held in a rubber like binding material. Hybrid engines primarily use a solid fuel and a liquid or gas oxidizer, allowing these engines to be shut down. Combining the simplicity of a solid rocket with the controllability (and safety) of liquid fuel engines has made hybrids attractive to many endeavours to lower launch costs including American Rocket's *Industrial Launch Vehicle*[14] and the winning X-prize contender Scaled Composites' *Spaceship One*.[15]

The air-breathing launch vehicles represent another gamble at lowering launch costs by reducing the need to carry onboard oxidizer. Elimination of oxidizer tanks would reduce launch mass, and vehicle bulk. However the need to intake and condition air for use by engines, whether by liquefying air as in the British HOTOL concept or by supersonic combustion ramjets (scramjets) results in heavy engines and vehicle design which force other compromises and problems in design to occur.[16] Proponents of pure rocket flight argue that these penalties surpass any benefits of not needing to carry large amounts of onboard oxidizer.[17] Ultimately the design of any launch vehicle is a matter of tradeoffs made in light of existing and achievable near term technology. The available materials at the time impact mass, as does the shapes used. Lockheed's famed Skunk Works was unable to produce a non-cylindrical composite fuel tank, contributing to the cancellation of Lockheed's X-33 space shuttle replacement testbed.[18] Wings and lift generating shapes impose penalties that add additional dead mass when not in use. Decisions concerning the use of stages or air launch must be considered against the costs of developing these additional components, along with operational costs of recovery and vehicle integration.

Even seeking low cost through reusability may be countered by the costs associated with refurbishment, as was found in the case of the US Space Shuttle. It is often claimed that building one vehicle that can be used several times makes more economic sense than building several vehicles that can be used only once. However from an operational perspective, the cost savings of not having to buy a launch vehicle for every launch must be balanced against the cost of reusing (or practically rebuilding) the launch vehicle. Returning a spacecraft to Earth after it has achieved orbit requires that the kinetic energy involved with an orbiting body be dissipated in a safe manner. A huge amount of energy is required to get things into orbit, and this same amount of energy is involved in getting down from orbit. The Earth's atmosphere is readily available to convert a spacecraft's kinetic energy into heat, but this is a dangerous process requiring a combination of careful manoeuvring and thermally protective materials. Features that allow for safe return: thermal protection, extra structure, flight control and landing gear all take up mass and bulk which could have been better used by payload. In this sense a reusable system will always be less efficient compared to an expendable launch vehicle built with the same technology. Once on the ground then there is the process of making ready the launch vehicle for reuse. The technology of the Space Shuttle requires a ground crew that is often described in terms of being a "standing army."[19] If the technology of the time does not allow for a cost-effective method of returning a launch vehicle to flight status, then there is no economic reason to design around reusability.

Launch vehicles have been designed since the Second World War; however, the design of a low-cost space launch capability remains elusive. The US Space Shuttle is the product of design tradeoffs made in light of the technology available at the time. As a low cost launch vehicle it has failed. The long and largely unsuccessful history of trying to replace the Space Shuttle only highlights the problems of technology not being able to live up to expectations. However without trying to reduce costs by actually funding launch vehicle design, then there is no way of knowing for certain if the technology is ready for cheap mass space access.

Space Launch by Gun

Gun launch of orbital payloads works somewhat differently in that the gun accelerates a payload in a very short period. In purely ballistic flight, the force of gravity will be decelerating the payload once it leaves the barrel. Rocket assistance is suggested to reduce the initial acceleration loads, though it is also possible to impart on a payload enough muzzle velocity so that its inertia will carry it completely to orbit. On the other hand, purely ballistic flight will not offer any opportunity for flight corrections if it was found that the forces resisting such flight (gravity, air resistance, wind speed etc.) were misjudged. Crewed payloads, such as Jules Verne's gun launched moon expedition are simply not possible, and computer controlled payloads must be especially hardy to survive accelerations measured in hundreds of thousands of times the force of gravity.[20] Chemical charges are only capable of accelerating a round so fast, therefore gun launch technologies include exotic light gas guns,[21] electromagnetic rail-guns,[22] and Electro Thermo Chemical (ETC) guns. While the latter two examples are also being examined as possible terrestrial mobile artillery, anything capable of space launch in the near term would be most likely be in a fixed position. Canadian Dr. Gerald Bull is associated with gun launch technology (and high powered artillery in general) prior to his assassination in 1990. Among his last projects was the incomplete Project Babylon Supergun for Iraq, which may have been part of a gun assisted space launch system.[23] The fixed nature of Bull's "superguns" and other guns capable of space launch limit application to launching large amounts of very small payloads into the same orbital plane.

Space Launch by Private Sector

In the spirit of New Public Management, there have been calls to reduce government's role in the space launch business, and allow private company's to freely compete in the hopes of creating a cost effective industry through competition. There would be, according to the proponents of this mode of space access, incentives for investors to fund risky technologies that would make space launch cheaper and even profitable. While the US lobby for this manner of space launch primarily targets what it views as wasteful spending by NASA, barriers to entrepreneurship, and the current arrangement of subsidized commercial launch by established big aerospace companies, the recent trend in outsourcing military logistics to private contractors may have military space launch as a potential untapped market for these entrepreneurs.

The development of economical space access is fraught with risks, and there are many competing technologies. The US, though the last remaining superpower, is unable to support all promising concepts. Indeed, proponents of private space access charge that the very way that government operates is hampering the whole development process. The recently won X-Prize competition did have as its benchmark a feat not yet matched by the US or other nation's space program: the repeated safe launch and recovery of the same passenger capable spacecraft to suborbital altitudes within a two week time limit.[24] The fast turnaround time and reusability conditions necessitates reliable and easy to maintain spacecraft technology. That such a capability was developed without public funds (another condition), only serves to embolden calls for more privatization of space access.

Demonstrations of private space flight and more importantly of a purely commercially viable orbital launch industry would change the space security and defence environment. X-Prize was an international competition, as is the ongoing struggle to create profitable orbital access. The fact that private concerns can even contemplate space access on their own is surely a sign of launch technology's proliferation. Despite potentially unlocking easy access to space, private investment may actually necessitate further government controls over space by making space control through arms a necessity. It is a paradox that a lowered bar for space access will give allies and adversaries increased ability to operate in space, reducing the advantaged position of the US in its military use of space.

Many companies have tried to enter into the launch industry with offers of alternative low-cost access only to fail. This is after all venture capitalism, with all the risks involved with developing and marketing technology. As long as these endeavours have a hint of profitability, then assuredly companies will be willing to challenge gravity and government regulation. Even without complete success, the private space access industry will have ramifications for the military use of space of all nations.

Micro Satellites

The "kill mechanism" for an attack against a spaceborne target is often less important than the platform doing the attacking. As long as it could be positioned properly, an open umbrella blocking a satellite's field of view could be considered a "kill mechanism." The difficult problem it would seem in many cases is to build a cheap enough satellite to take a disruptive or damaging payload within effective range of a target. The micro-satellite concept is perhaps the most promising of technologies that will give the US or others such a capability.

Due to being smaller, a micro, nano or pico prefixed satellite is individually less capable, often performing only a single task and for a shorter lifespan. This can be argued as being a disadvantage. On the other hand this opens up opportunities for cost savings over larger satellites:

1. A small short lived satellite requires frequent replacement; this implies that upgraded technologies can be progressively added as they

become mature. In the event that a design or concept deficiency is found, then there is no technical reason these cannot be corrected for later satellites sent up as replacements.

2. Short life spans and regular replacement can translate into lower engineering standards. There is reduced need for near perfection or "indestructibility" as these spacecrafts are better compared with munitions as opposed to warships in terms of cost and complexity. Similarly, an opponent's counter space plans will have to contend with large numbers of targets[25] instead of a handful of technological "masterpieces."

3. The current interest in open architectures almost certainly means that a few small spacecraft "families" based on common components and design will emerge. This is part of the criticism of the XSS Micro-satellite program, as it is both proposed (and feared depending on ones perspective in the space weaponization debate) that one of the designs in the XSS series will be adapted into an ASAT weapon.[26] Mass production of basic satellite components would also promise cost reductions.

4. Given that current and foreseeable launch costs are prohibitive, there is good reason to make space systems as mass efficient as possible. Smaller payloads potentially open up the use of lower cost small payload launch technologies, such as air launched rockets, which are difficult to scale up within available technology.

5. It has been argued by those concerned with a viable orbit access industry that increased demand for launch vehicles will encourage the development of low cost spaceflight.[27]

It must be remembered that any benefits micro-satellite and other space technology geared towards lowering the cost of space may translate into a threat if the technology spreads to hostile interests. This is a global economy, and the proliferation of economically useful technologies is part of it. Unlike weapons of mass destruction, the technology needed for micro-satellites, cluster computers and enhanced space access have direct civilian and private sector applications. Therefore, unlike the slow proliferation of WMD, these and other technologies critical to lowering the costs of space militarization will undoubted spread quickly. For these reasons, technologies that have an economical benefit to arguments made in favour space weaponization, may actually have a secondary effect of encouraging weaponization through the creation and enhancement of threats.

Notes

1. It is of note that the inclusion or exclusion of attacks against the earthbound infrastructure and client base of space services in the definition of space warfare is a hotly contested part of the contemporary debate over space weaponization.
2. Ames Research Center, National Aeronautics and Space Administration. "Remote Agent Experiment," <http://ic.arc.nasa.gov/projects/remote-agent/faq.html> (2005).
3. Matthew Mowthorpe, *The Militarization and Weaponization of Space*, (Toronto: Lexington Books, 2004) 22.
4. Institute of Electrical and Electronics Engineers, Taskforce on Cluster Computing. "IEEE Taskforce on Cluster Computing," available from <http://www.ieeetfcc.org> (2005).
5. SAR type systems are capable of obtaining image resolutions rivalling that of optical systems, but are less affected by weather.
6. Dawn Stover, "The New War in Space," *Popular Science*, September 2002, 43.
7. Such a proposal would also require computers capable of filtering contacts as global coverage of all discernable heat sources (let alone generating images based on combining thermal images) would be an overwhelming amount of data.
8. Tim Folger, "Shield of Dreams," *Discover* 22 no. 11 (November 2001): 62.
9. Attack by Direct Energy Weapons would be an entirely different matter. Potentially the best a defender against a DEW attack would be able to accomplish is to identify that an attack had taken place.
10. Matthew Mowthorpe, *The Militarization and Weaponization of Space*, 121.
11. Collins, John M. *Military Space Forces*. (Washington: Pergamon-Brasesey's International Defense Publishers, Inc., 1989), 16.
12. The US Saturn V launch vehicle stands out a one of the few examples of launch vehicles that have not suffer a single catastrophic launch failure. This safety record is all the more impressive in light of its role in the US victory in the Moon Race. On the other hand the Saturn V's Soviet era competition, the similarly powerful N-1 launch vehicle was never able to get off the launch pad without blowing up.
13. Preston Lerner, "A few Dreamers Building Rockets in Work-Shops," *Popular Science*, May 2003, 59.
14. Global Security. "Industrial Launch Vehicle / Aquila / American Rocket [Amroc]," <http://www.globalsecurity.org/space/systems/amroc.htm.htm> (2005).
15. Scaled Composites. "Tier One Private Manned Space Program," <http://www.scaled.com/projects/tierone/faq.htm> (2005).
16. With scramjets, not only must the airframe be it be capable of hypersonic atmospheric flight, but it also must be part of the engine, and still be lightweight enough to carry a worthwhile payload.
17. Andrew J. Butrica, *Single Stage to Orbit*, (Baltimore: The Johns Hopkins University Press, 2003), 81.
18. Ibid, 210.
19. Ibid, 68.

20. Thomas R. McDonough, *Space The Next Twenty-Five Years*, Revised and Updated Edition (Toronto: John Wiley & Sons ,Inc, 1989), 186.

21. The "light gas" of a light gas gun refers to the use of a gas made up of molecules with a low mass, such helium or hydrogen. Light gases can be mechanically accelerated (compression by explosively driven pistons) to velocities in excess of those imparted by detonation on combustion products (heavier complex molecules such as water) alone.

22. Another form of linear motor assisted space launch involves the use of magnetically levitated and driven sleds throwing launch vehicles into the air. With a long enough track, or lower release velocities sled assisted launch would be tolerable by flight crews.

23. Global Security. "Project Babylon Supergun / PC-2," <http://www.globalsecurity.org/wmd/world/iraq/supergun.htm> (2005).

24. X Prize Foundation. "What is the X PRIZE™?" <http://www.xprize.org/press/what.html> (2004).

25. This however does bring up a strong rational for using mass destruction space denial techniques.

26. Jeffrey Lewis, "Space Weapons in US Defense Planning." *INESAP Information Bulletin* April 2004. <http://www.inesap.org/bulletin23/art03.htm> (2005).

27. Greg Klerkx, *Lost in Space: The Fall of NASA and The Dream of a New Space Age*, (New York: Pantheon Books, 2004), 103.

APPENDIX TWO

Revenge of the Lighter-than-Air Craft

Labeling outer space as the "highest of high-grounds," for many only brings up the matter of finding a more modest "high-ground," suitable for many of the same tasks. There is a wide band of atmosphere between the operational ceilings of conventional airplanes and the lowest possible altitudes for orbit. At these so called "near-space"[1] altitudes the thin atmosphere makes aerodynamic lift difficult to generate, yet is still too dense to allow for orbit.[2] Interest in this part of the atmosphere and new technologies are now allowing the oldest of flight technologies, lighter-than-air platforms, to compete against sophisticated high-altitude airplanes in this relatively unexploited environment. Commercial and military interests see promise in high-altitude lighter-than-air technology to supplement and perhaps even replace some orbiting satellites.

For its opponents, lighter-than-air technology seems to be historical anachronisms, which fell out of fashion decades ago. The problems of control and basic fragility have spelled the end of many airships. The tragic histories of the two US Navy aircraft carrying airships U.S.S. Macon and U.S.S. Akron, the British airship R101, and most infamously, the German Hindenburg[3] all seem to warn of the dangers found in airship design. Though aircrafts have a natural tendency to seek out the ground when fuel supplies ran out, powered flight has come to dominate air power. Airship technology was to a large degree superseded during the early 20th century in favor of increasingly more capable aircraft technology.

There is some element of truth to lighter-than-air platforms simply falling out of fashion.[4] A stately airship is not what immediately comes to mind when one thinks of aerospace power. However as the realm of telecommunications (and specifically broadband) have demonstrated, less glamorous Earth bound technologies such as fiber optics can triumph over more exciting technologies such as the communication satellite. Cost effectiveness is usually a greater force than enthusiasm for the fashionable. The rigid airships of the early 20th century suffered from a set of costs, performance and vulnerabilities which made them uncompetitive against heavier-than-air vehicles for the airpower missions of strategic bombing, cargo,[5] and as the USN attempted, power projection. Indeed the airship has never entirely disappeared as its strengths such as persistence and stealthy flight[6] have found it niche roles. In the debate over space power, lighter-than-air craft performance at extreme altitudes potentially offer a winning set of capabilities verses costs when compared to satellites.

Altitude above a battlefield allows for greater line-of-sight for communications, surveillance and targeting. Height in these terms equals information superiority. Height therefore is a militarily important asset. The orbiting satellite is perhaps the ultimate expression of the imperative to be above it all. Launch vehicles, burn propellant loads that dwarf the payloads actually delivered to orbit. The rewards of this expensive technology are the globally commanding heights of orbit, the "highest of high ground." Arguments made in favour of rapid and aggressive space weaponization go so far as to argue that space is ultimately the strategic realm of most consequence.

Strategic capabilities are not always useful to the tactical battlefield, in that they often lack flexibility and responsiveness to accommodate the needs of those immediately in harms way. While there is a very strong case for the US in the 1991 Gulf War having its military, "center of gravity," located in space,[7] it can also be strongly argued that this war identified severe problems with getting the benefits of US space power down to the tactical level of those immediately in the war zone. Satellite resources while influencing all US and allied forces in the 1991 conflict, were not available to all who could have potentially used them. Simply put there were and still are not enough satellites to go around.

All levels of the military seem to want more communications bandwidth, overhead reconnaissance and targeting information. Later wars, such as ongoing operations in Afghanistan and Iraq, have seen marked progress in the proliferation of satellite based capabilities. Despite charges that the US is in some ways flaunting its information advantage (there are questions as to the military importance of video conferencing);[8] demand grows, straining contemporary capacity. The increasing use of commercial satellite resources to supplement purely military assets is symptomatic of the great demand for space supplied information superiority.

Present day satellites are for the most part expensive, few in numbers and largely locked into predictable orbits. The US lead in space militarization has the side effect of producing an asymmetrical target in these very space systems. The more satellites are used in conventional warfare; arguably the more tempting it becomes to target them. Warnings of a potential "Space Pearl Harbor,"[9] highlight such fears in the US. Deterrence, arms control and defensive weapons are all proposed as solutions to perceived satellite vulnerabilities. While various combinations of these measures may produce some degree of security in space, a comprehensive plan includes having redundancies and alternatives.

The "effect" of an information edge is not platform dependent. In a cost conscious spending climate, if expensive satellites could be replaced by a lower cost solution it almost certainly would be. As part of examining the current and future state of space militarization, there can be identified several alternative technologies to the present day US space infrastructure. These include changing operational concepts to smaller more easily mass produced/launched micro satellites, and high atmospheric aircrafts. Lighter-than-air platforms stand out due to their capability to operate at extreme heights cheaply. Low-cost implies the ability to procure more assets which in turn allows more capacity and importantly gives tactical end users more control over the assets providing their information. In addition airships and balloons are arguably less vulnerable to attack due to being harder to detect and observe. Increasing the inherent "stealthiness" of an airship can be done at relatively low cost, which is in contrast with practically all other platforms. Responsiveness, flexibility and survivability are characteristic being perused by US and other modern military forces.

Lighter-than-Air vehicles get most of their lift for "free." Instead of harnessing lift generated by airflow over wings or by direct application of thrust, pure lighter-than-air craft uses simple buoyancy for its movement and persistence in the vertical realm. If station keeping is a desired trait, then engines may be added as in the case of blimps and rigid frame airships. The propulsion required to actively position an airship is visibly less energetic than that needed for a conventional powered heavier-than-air craft of similar payload.

A problem of lighter-than-air systems is that in addition to providing lift, a lighter-than-air craft's lifting envelope also acts as a sail. The comparatively lightweight propulsion systems found on early dirigibles were often found to be lacking when flying against strong winds, let alone severe weather. This is the reality of flight at lower altitudes. At the extreme altitudes being proposed, platforms will be far above the weather, and be facing relatively calm wind conditions.[10] As fragile and difficult to steer as airships are at lower altitudes, these problems pale in comparison to the dangers and preparation needed for lofting the average satellite with contemporary launch systems.

Advances in materials, electric propulsion, generation and storage, in conjunction with satellite assisted navigation[11] are critical to allowing the old technology of airships to compete against satellite constellations for some missions. In applications such as data relay or surveillance, staying within range is the only movement of concern. For a pure airship maintaining altitude is free,[12] the only propulsion necessary is for station keeping and perhaps deployment. The mechanics of orbiting mean that aside from satellites in GeoEarth orbits, most satellites cannot be always overhead. To provide continuous coverage of a battlefield, large constellations are needed, ranging from roughly two dozen for higher orbiting GPS satellites, to hundreds proposed for the LEO communications satellite networks of the 1990's, and over a thousand for the proposed Brilliant Pebble missile defence system.[13] A few high altitude dirigibles can in rotation provide constant coverage over a small region for indefinite periods.

Alternatively balloons released like sonar buoys at sea could provide similar capabilities. In an age of reasonably accurate weather monitoring and disposable electronics, unguided single-use platforms could prove for some applications to be more economic than constructing maneuvering systems. Even with un-powered balloons there are practical ways recover equipment. The utility of balloon lofted payloads have recently been tested under the US Combat SkySat which involved the use of commercially developed balloons and existing military radio equipment to extend tactical communication ranges.[14]

If more of a service was needed, then it would be necessary to move additional platforms into theatre. The comparatively low cost of a dirigible allows for greater numbers to be procured. Deploying additional airships to a theatre only requires a fraction of the resources needed to loft a satellite. Conceivably an airship could be modular, with different mission specific modules being added as needed further reducing costs.[15] An equivalent space based infrastructure is not as flexible due to the cost of procurement, launch and orbital management of satellites. Long duration satellites, as currently favoured by the US, tend to be more expensive due to the need for robust equipment capable of operating for years without servicing. While there is nothing to stop spacecraft architecture from being modular or serviceable, there are the problems of either swapping out components in orbit, or reentry for changing modules on the ground. Interestingly, proposals for reconfigurable spacecraft such as the self-landing Space Maneuver Vehicle[16] or autonomous on-orbit servicing by micro-satellites have faced controversy over their weapons potential.[17]

True lighter-than-air craft by definition, must stay within the atmosphere it floats in. This limits the maximum altitude of balloons and airships to well below that of satellites. Height means further horizons (longer line-of-sight ranges), but it also means being further away from important points on the ground. This is a trade-off between coverage area and distance to points of interest on the ground. For different missions there is a different balance between these concerns. Global systems for obvious reasons must emphasize range above all else. High orbit satellites, give commanding views of almost entire hemispheres, but may not offer enough focus or responsiveness for battlefield applications. Lower orbit systems, such as the commercial LEO communications satellite projects of the 1990's made up for smaller satellite coverage areas through sheer numbers. Both large LEO constellations and high altitude orbit satellites, with present technology, are expensive propositions.

In their defence, GPS, the mass LEO satellite constellations, and Brilliant Pebble type proposals are meant to provide their services across the entire globe without interruption. Satellite constellations are uniquely suited for global coverage and wide fields of view. Lighter-than-air craft and high flying aircraft can only provide local coverage. Global coverage by high altitude or near-space airships would require even more platforms than those balked at for the LEO constellations. Thus far the comparison is between lighter-than-air platforms verses satellite constellations for providing effective battlefield coverage of services necessary for local information supremacy. That is to say that there is nothing for breakthroughs in satellite and space launch technology to tilt the cost to benefits equation in favour of purely space based solutions. Similarly not all space systems appear to have a direct substitute within the near-space capabilities now being pursued.

With present and near term technology however, the cost verses capabilities ratio seems to favour the high-atmospheric craft in the provision of services to subscribers (or targets) located in a fixed service areas. At present commercial, scientific and military communities already have or are developing lighter-than-air capabilities. The current technology in use by NASA has achieved flight times measured in weeks, with plans for even longer durations.[18] In the private sector, there are investors willing to bet on using a small number of high altitude airships to replace dense networks of less capable cellular phone ground stations in urban areas.[19] This is the commercial application of the greater lines-of-sight provided by high-altitude, but without the costs to both subscribers and service provider of two-way satellite communications.

The characteristics that have made high altitude lighter-than-air technology attractive to the scientific and private sectors are also of value to the military. Exploitation of these high, but not orbital high altitudes, has been terms as being, "a low-threat, high-payoff environment."[20] The current rosy perception of near space contrasts with pessimistic arguments that space is a potential "Achilles Heel," in addition to clearly being very expensive.

In the debate over how to deal with the emerging threat to Western space systems, a viable low-cost satellite replacement potentially could make some parts of the debate seem over blown. Mass deployment of near-space systems would tend to reduce the chance that an attack against US space infrastructure, even one involving mass destruction of whole constellations would be a "knock out blow" to the US. In effect a "space Pearl Harbour" would still allow the US to continue the fight on with other information superiority systems. This would parallel the source of the metaphor in that the loss of US Pacific based battleships in 1941 was made up for by the submarine and carrier aviation branches of the US Navy. Perhaps the nightmare scenario of mass space attack will find dirigibles/high-flying aircraft pseudo satellites, ground based navigation aides, and other non-space information superiority resources filling in to allow Western forces to prevail.

That is not to say that all tasks can be accomplished from within the atmosphere. As mentioned before, high-attitude balloons and airships are only envisioned as theatre systems, and cannot offer the global persistence that only a satellite constellation can. Loss of space infrastructure, even with pseudo-satellite stand-ins would be a severe blow to US global capabilities. The existence of an effective high-atmospheric substitute would however introduce additional doubt over victory for the threat nation that somehow destroyed US space superiority. Doubt of certain victory through space denial may deter some threats from pursuing such a strategy, but perhaps not all.

Before concluding, there is one final use for high altitude lighter-than-air craft that is of interest to the debate over space weaponization, assisting in space launch. Several contenders for the X-Prize private suborbital space launch competition had planned to use lighter-than-air technology to lift their spacecraft to altitude. Instead of using powered flight, contenders such as the Canadian DaVinchi Project[21] planned to use balloons to lift of their manned sub-orbital rockets to launch altitude. Smaller payloads on more powerful rockets can make use of these same balloons to assist in reaching orbit. A more ambitious plan is proposed by JP Aerospace, where technology willing, an airship capable only of survival in the thin atmosphere of near-space would slowly climb from a platform already in near-space and accelerate to low Earth orbit via efficient electric propulsion.[22] As far fetched as this manner of (some would say ludicrously) subtle space launch seems, JP Aerospace is also demonstrating autonomous airship technology for the US military as part of the current interest in using more near term technology as satellite substitutes.[23]

Altitude above a battlefield has historically always bestowed an information superiority potential. Though space is the "highest of high grounds," the costs involved in using this vantage point are currently great. Indeed, such extreme heights may even act as a disadvantage due to unnecessary distance between platform and points of interest underneath. For this reason there is now interest in lighter-than-air platforms for to complement space based platforms. Developments in this field of aviation may offer serious alternatives to many near term space militarization and weaponization capabilities. Supplanted by powered flight early in the 20th century, lighter-than-air craft technology may ultimately limit near term application of orbital space and the powerful means for getting there.

Notes

1 Hampton Stephens, "Near-Space." *Air Force Magazine Online* 88, no 7, (July 2005). <http://www.afa.org/magazine/July2005/0705near.asp> (2005).

2 In addition, gravity's pull towards the Earth is stronger, requiring even higher speeds to maintain orbit if the atmosphere was not present.

3 It should be noted that the US airships were filled with inert helium, while the two named European airships were lifted by combustible hydrogen at the time of their ends. During the inter-war period, the US controlled most of the world's helium supply.

4 Hampton Stephens, "Near-Space." *Air Force Magazine Online* 88, no 7, (July 2005). <http://www.afa.org/magazine/July2005/0705near.asp> (2005).

5 There is recent interest within the transport industry and military in taking advantage of dirigible technology's inherent low-cost and developing ability to go practically anywhere for cargo carrying.

6 Airships are reasonably quite due to low propulsion requirements. Similarly the low power propulsion does not present much of a heat signature. Technology can further mask both acoustic and heat signatures. Lifting envelopes can be made of radar transparent materials, while ancillary systems and payloads can make use of shaping and radar absorbent materials to avoid radar detection (stealth technology).

7 George Friedman, and Meredith Friedman, *The Future of War*, (New York: St. Martin's Griffin, 1996) 303.

8 Michael E. O'Hanlon, *Neither Star Wars Nor Sanctuary*, (Washington, DC: Brookings, Institution Press, 2004), 126.

9 The Commission to Assess United States National Security Space Management and Organization, *Report of The Commission to Assess United States National Security Space Management and Organization*, 11 January 2001. <http://www.space.gov/docs/fullreport.pdf> (2004).

10 Global Security. "High Altitude Airship," <http://www.globalsecurity.org/military/systems/munitions/bgm-109-var.htm> (2005).

11 The utility of global satellite navigation and communication for this potential satellite supplement means that lighter-than-air craft are not a total replacement for critical military satellites.

12 Hybrid airships with lifting envelopes shaped to generate aerodynamic lift at speed are also proposed. These require some movement forward to maintain flight.

13 Matthew Mowthorpe, *The Militarization and Weaponization of Space*, (Toronto: Lexington Books, 2004) 22.

14 Hampton Stephens, "Near-Space." *Air Force Magazine Online* 88, no 7, (July 2005). <http://www.afa.org/magazine/July2005/0705near.asp> (2005).

15 With a high altitude balloon, changing mission equipment is often a matter of hanging a different box.

16 Dawn Stover, "The New War in Space," *Popular Science* 261, no 3, (September 2002): 47

17 Current controversies parallel similar Cold War concerns over the potential for the US Space Shuttle to have both reconfigurable mission payloads and allow for on-orbit servicing. Then as now both capabilities were viewed as being potentially weaponizable.

18 National Aeronautics and Space Administration. "Balloon Program Office," <http://www.wff.nasa.gov/~code820/missions/missions.html> (2005).

19 Xeni Jardin, "Bird? Plane? UFO? No, Stratellite." *Wired News*, 23 December 2002. <http://www.wired.com/news/wireless/0,1382,56961,00.html> (2005).

20 Hampton Stephens, "Near-Space." *Air Force Magazine Online* 88, no 7, (July 2005). <http://www.afa.org/magazine/July2005/0705near.asp> (2005).

21 Da Vinci Project. "Worlds Largest Reusable Helium Balloon Completed by

Canadian Manned Space Flight Team," 2 December 2004. <http://www.davinciproject.com/beta/News/NewsMain.html> (2005).

22 JP Aerospace. "Airship to Orbit," available at <http://www.jpaerospace.com/atohandout.pdf> (2005).

23 Michael Sirak, "US Air Force eyes 'near space' vehicle." *Jane's Defence Weekly* 19 September 2003.
<http://www.janes.com/defence/news/jdw/jdw030919_1_n.shtml> (2005).

References

Primary Sources

Ames Research Center, National Aeronautics and Space Administration. "Remote Agent Experiment." <http://ic.arc.nasa.gov/projects/remote-agent/faq.html> (2005).

The Commission to Assess United States National Security Space Management and Organization. *Report of The Commission to Assess United States National Security Space Management and Organization*, 11 January 2001. <http://www.space.gov/docs/fullreport.pdf> (2004).

Department of Defense. *Findings of the Nuclear Posture Review*, 9 January 2002. <http://www.defenselink.mil/news/Jan2002/020109-D-6570C-001.pdf> (2004).

Department of State. "Treaty on Principles Governing the Activities of States in the Exploration and Use of Outer Space, Including the Moon and Other Celestial Bodies," 27 January 1967. <http://www.state.gov/t/ac/trt/5181.htm> (2004).

National Aeronautics and Space Administration. "Balloon Program Office." <http://www.wff.nasa.gov/~code820/missions/missions.html> (2005).

National Aeronautics and Space Administration. "Earth's Atmosphere," 1 December 1995. <http://liftoff.msfc.nasa.gov/academy/space/atmosphere.html> (2004).

National Aeronautics and Space Administration, "Orbital Velocity and Period Calculator," 23 June 1995. <http://liftoff.msfc.nasa.gov/academy/space/atmosphere.html> (2005).

National Aeronautics and Space Administration. "Skylab Operations Summary," 29 September 2000. <http://www-pao.ksc.nasa.gov/kscpao/history/skylab/skylab-operations.htm> (2004).

United Nations. "GA/SPD/192: RUSSIAN FEDERATION CAUTIONS AGAINST MILITARY DEPLOYMENT IN OUTER SPACE; REITERATES PROPOSAL FOR CONFERENCE TO PREVENT MILITARIZATION," 17 October 2000. <http://www.un.org> (2004).

United States Air Force. *Air Force Doctrine Document 1*, 17 November 2003. <https://www.doctrine.af.mil/Library/Doctrine/afdd1.pdf> (2004).

United States Air Force. *Air Force Doctrine Document 2-2*, 27 November 2001. <https://www.doctrine.af.mil/Library/Doctrine/afdd2-2.pdf> (2004).

United States Air Force. *Air Force Doctrine Document 2-2.1*, 2 August 2004. <http://www.dtic.mil/doctrine/jel/service_pubs/afdd2_2_1.pdf> (2004).

United States Strategic Command. *Space Missions*, March 2004. <http://www.stratcom.mil/factsheetshtml/spacemissions.htm> (2004).

Secondary Sources

21st Century Airships. "About 21st Century Airships." <http://www.21stcenturyairships.com/AboutUs> (2005).

Barker, Kenneth W. "Airborne and Space-Based Lasers." In *The Technological Arsenal*, ed. William C. Martel, 38-54. Washington: Smithsonian Institution Press, 2001.

Belote, Howard D., Major, USAF. "The Weaponization of Space." *Airpower Journal* (Spring 2000). <http://www.airpower.maxwell.af.mil/airchronicles/apj/apj00/spr00/belote.htm> (2004).

Berkowitz, Bruce. *The New Face of War*. Toronto: The Free Press, 2003.

Biddle, Tami Davis. *Rhetoric and Reality in Air Warfare*. Princincton: Princeton University Press, 2002.

Boeing, "XSS Micro-satellite." <http://www.boeing.com/defense-space/space/xss/> (2004).

Boot, Max. "The New American Way of War." *Foreign Affairs* 82, no 4 (July/August 2003). <http://www.foreignaffairs.org/20030701faessay15404/max-boot/the-new-american-way-of-war.html> (2004).

British Broadcasting Corporation. "China joins EU's satellite network," 19 September 2003. <http://news.bbc.co.uk/2/hi/science/nature/3416231.stm> (2004).

British Broadcasting Corporation. "Brazil vows to pursue space plan," 23 August 2003. <http://news.bbc.co.uk/1/hi/world/americas/3176395.stm> (2004).

British Broadcasting Corporation. "Mir Space Station 1986-2001." <http://news.bbc.co.uk/hi/english/static/in_depth/sci_tech/2001/mir/default.stm> (2004).

British Broadcasting Corporation. "SpaceShipOne rockets to success," 4 October 2004. <http://news.bbc.co.uk/1/hi/sci/tech/3712998.stm> (2004).

Butrica, Andrew J. *Single Stage to Orbit*. Baltimore: The Johns Hopkins University Press, 2003.

Chipman, Donald D., Dr. "AIRPOWER A New Way of Warfare (Sea Control)." *Airpower Journal*, Fall 1997. <http://www.airpower.maxwell.af.mil/airchronicles/apj/apj97/fal97/chipman.html> (2004).

Clancy, Tom and Chuck Horner, General, USAF(retired) *Every Man a Tiger*. New York: G.P. Putnam's Sons, 1999.

Clausewitz, Carl von. *On War*. Everyman's Library Ed. Translated and Edited. Michael Howard and Peter Paret. Princeton University Press, 1976. Reprint Toronto: Alfred A. Knopf, 1993.

Collins, John M. *Military Space Forces*. Washington: Pergamon-Brasesey's International Defense Publishers, Inc., 1989.

Cornwell, Bernard. *Historical Note*. In *Sharpe's Escape*. London: HarperCollins, 2004.

Da Vinci Project. "Worlds Largest Reusable Helium Balloon Completed by Canadian Manned Space Flight Team," 2 December 2004. <http://www.davinciproject.com/beta/News/NewsMain.html> (2005).

Deblois, Bruce M., Lt. Col., USAF, "Space Sanctuary A Viable National Strategy," *Aerospace Power Journal* (Winter 1998). <http://www.airpower.maxwell.af.mil/airchronicles/apj/apj98/win98/deblois.html> (2004).

DeBlois, Bruce M., Richard L. Garwin, R. Scott Kemp, and Jeremy C. Marwell. "Space Weapons: Crossing the U.S. Rubicon." *International Security* (Fall 2004). <http://mitpress.mit.edu/catalog/item/default.asp?ttype=4&tid=26> (2005).

DeKok, Roger G. and Bob Preston, "Acquisition of Space Power for the New Millennium." In *Spacepower for a New Millennium*, ed. Peter L. Hays, James M. Smith, Alan R. Van Tassel, and Guy M. Walsh, 61-90. New York: McGraw-Hill, 2000.

Dolman, Everett C. *Astropolitik*. Portland: Frank Cass, 2002.

Dupont, Daniel G. "Nuclear Explosions in Orbit." *Scientific American* 290 no. 6 (July 2004): 100-107.

The Eisenhower Institute. "A European Perspective on Current Trends in Military and Civilian Space," 2004. <http://www.eisenhowerinstitute.org/programs/globalpartnerships/fos/newfrontier/parismeeting.htm> (2005).

"Eye spy." *The Economist*. Nov 10, 2001. Quoted in Global Security. <http://www.globalsecurity.org/org/news/2001/011110-eye.htm> (2004).

Fédération Aéronautique Internationale. "100 km Boundary for Astronautics," 25 June 2004. <http://www.fai.org/book/view/22> (2004).

Federation of American Scientists. "FAS Calls for Alternatives to Weapons in Space." 8 October 2004. <http://fas.org/nuke/control/os/> (2004).

Folger, Tim. "Shield of Dreams." *Discover* 22 no. 11 (November 2001): 58-67.

France, Martin E. B., Lt Colonel. "Back to the Future: Space Power Theory and A. T. Mahan." *Air & Space Power Chronicles*, August 2000. <http://www.airpower.maxwell.af.mil/airchronicles/cc/france1.html> (2004).

France, Martin E. B., Lt Colonel. "Planetary Defense: Eliminating the Giggle Factor." Air & Space Power Chronicles, August 2000. <http://www.airpower.maxwell.af.mil/airchronicles/cc/france2.html> (2004).

Friedman, George and Meredith Friedman. *The Future of War*. New York: St. Martin's Griffin, 1996.

Global Security. "BGM-109 Tomahawk." <http://www.globalsecurity.org/military/systems/munitions/bgm-109-var.htm> (2004).

Global Security. "High Altitude Airship." <http://www.globalsecurity.org/military/systems/munitions/bgm-109-var.htm> (2005).

Global Security. "Navstar Global Positioning System." <http://www.globalsecurity.org/space/systems/gps.htm> (2005).

Global Security. "Nodong-1." <http://www.globalsecurity.org/wmd/world/dprk/nd-1.htm> (2004).

Global Security. "Project Babylon Supergun / PC-2." <http://www.globalsecurity.org/wmd/world/iraq/supergun.htm> (2005).

Godwin, Robert, ed. *Dyna-Soar Hypersonic Strategic Weapons System.* Burlington: Apogee Books, 2003.

Grossman, Lev. "Beyond the Rubber Bullet." *Time Online Edition* 21 July 2002. <http://www.time.com/time/nation/article/0,8599,322588,00.html> (2004).

Hays, Peter L., James M. Smith, Alan R. Van Tassel, and Guy M. Walsh, ed. *Spacepower for a New Millennium*, ed. New York: McGraw-Hill, 2000.

Hitchens, Theresa. "Europe's USAF Counterspace Operation Doctrine: Questions Answered, Questioned Raised." *Center for Defense Information*, 4 October 2004. <http://www.cdi.org/program/document.cfm?DocumentID=2504&from_page=../index.cfm> (2004).

Hobbs, David. *Space Warfare.* New York: Prentice Hall, 1986.

Institute of Electrical and Electronics Engineers, Taskforce on Cluster Computing. "IEEE Taskforce on Cluster Computing." <http://www.ieeetfcc.org/> (2005).

International Campaign to Ban Landmines. "Treaty Members." 21 July 204. <http://www.icbl.org/treaty/members?eZSESSIDicbl=5076874f9a019e800212f27d01202eba> (2004).

Jardin, Xeni. "Bird? Plane? UFO? No, Stratellite." Wired News, 23 December 2002. <http://www.wired.com/news/wireless/0,1382,56961,00.html> (2005).

JP Aerospace. "Airship to Orbit." <http://www.jpaerospace.com/atohandout.pdf> (2005).

Kitts, Christopher A. and Richard A. Lu. "The Stanford SQUIRT Micro Satellite Program." 7 June 7 1994. <http://ssdl.stanford.edu/aa/papers/SSDL9404.pdf> (2004).

Klerkx, Greg. *Lost in Space: The Fall of NASA and The Dream of a New Space Age.* New York: Pantheon Books, 2004.

Krepon, Michael, Jeffery Lewis, and Theresa Hitchens. "Weapons in Space."

Arms Control Today November 2004. <http://www.armscontrol.org/act/2004_11/Krepon.asp> (2004).

Lambeth, Benjamin S. *Mastering the Ultimate High Ground: Next Steps in the Military Uses of Space*. RAND, 2003. <http://www.rand.org/publications/MR/MR1649/> (2004).

Lawrence Livermore National Laboratory. "HyperSoar." <http://www.llnl.gov/str/Carter.html> (2004).

Lerner, Preston. "A few Dreamers Building Rockets in Work-Shops." *Popular Science* 262, no 5, (May 2003): 56-64.

Lewis, Jeffrey. "Space Weapons in US Defense Planning." *INESAP Information Bulletin* April 2004. <http://www.inesap.org/bulletin23/art03.htm> (2005).

Lockwood, Jonathan S. "Space Control Versus Space Denial in 21st Century Warfare: Achilles Heel of the RMA (Revolution in Military Affairs)?" *Defense & Foreign Affairs Strategic Policy 28*, no.8 (2000):4.

Logsdon, John M. "Just Say Wait to Space Power." *Issues In Science and Technology 17*, no 3, (Spring 2001). <http://www.issues.org/17.3/p_logsdon.htm> (2004).

Lupton, David E. Lt. Colonel (Retired), USAF. *On Space Warfare*. Maxwell Air Force Base, Alabama: Air University Press, 1998. <http://www.airpower.maxwell.af.mil/airchronicles/apj/apj98/win98/deblois.html> (2004).

McDonough, Thomas R. *Space The Next Twenty-Five Years*, Revised and Updated Edition. Toronto: John Wiley & Sons ,Inc, 1989.

McElyea, Tim. *A Vision of Future Space Transportation*. Burlington: Apogee Books, 2003.

McKinley, Cynthia A.S, Lt. Colonel, USAF. "The Guardians of Space." *Airpower Journal*, Spring 2000. <http://www.airpower.maxwell.af.mil/airchronicles/apj/apj00/spr00/mckinley.htm> (2004).

McKitrick, Jeffrey, James Blackwell, Fred Littlepage, George Kraus, Richard Blanchfield and Dale Hill. "The Revolution in Military Affairs." *Battlefield of the Future*. <http://www.airpower.maxwell.af.mil/airchronicles/battle/ov-4.html> (2004).

Morton, Oliver. "Europe's New Air War." *Wired Magazine*, August 2002. <http://wired.com./wired/archive/10.08/airwar.html> (2004).

Mowthorpe, Matthew. *The Militarization and Weaponization of Space*. Toronto: Lexington Books, 2004.

Mueller,Karl P. "Totem and Taboo: Depolarizing the Space Weaponization Debate," (Paper based on presentation given to Weaponization of Space Project of the Eliot School of International Affairs Space Policy Institute and Security Policy Studies Program, George Washington

University, 3 December 2001). <http://www.gwu.edu/~spi/spaceforum/TotemandTabooGWUpaperRevised%5B1%5D.pdf> (2004).

Nguyen, Hung. "Russia's Continuing Work on Space Forces," *Orbits*, Summer 1994, 413-423. Quoted in Matthew Mowthorpe. *The Militarization and Weaponization of Space*, 70. Toronto: Lexington Books, 2004.

Oberg, James. "The war of words over war in space," 16 April 2004. <http://msnbc.msn.com/id/4732874> (2004).

O'Hanlon, Michael E. *Neither Star Wars Nor Sanctuary*. Washington, DC: Brookings, Institution Press, 2004.

Reynolds, Clark G. *Command of the Sea*. New York: William Morrow & Company, Inc., 1974.

Rife, Shawn P., Major, USAF. "On Space-Power Separatism," *Airpower Journal*, Spring 1999. <http://www.airpower.maxwell.af.mil/airchronicles/apj/apj99/spr99/rife.html> (2004).

Scaled Composites. "Tier One Private Manned Space Program." <http://www.scaled.com/projects/tierone/faq.htm> (2005).

Singer, Jeremy. "Satellite Jammer Ready: U.S. Parallel Effort To Thwart Imaging Craft Dropped." C4ISR Journal 19 October 2004. <http://www.c4isrjournal.com/story.php?F=461040> (2004).

Sirak, Michael. "US Air Force eyes 'near space' vehicle." *Jane's Defence Weekly 19* September 2003. <http://www.janes.com/defence/news/jdw/jdw030919_1_n.shtml> (2005).

Smith, Bob. "The Challenge of Space Power." *Aerospace Power Journal*, Spring 1999. <http://www.airpower.maxwell.af.mil/airchronicles/apj/apj99/spr99/smith.html> (2004).

Stares, Paul B. *Space Weapons & US Strategy: Origins & Development*. London: Croom Helm, 1985.

Stephens, Hampton. "Near-Space." *Air Force Magazine Online 88*, no 7, (July 2005). <http://www.afa.org/magazine/July2005/0705near.asp> (2005).

Stover, Dawn. "The New War in Space." *Popular Science* 261, no 3, (September 2002): 40-47.

Sweetman, Bill. "Space Shuttle: The Next Generation." *Popular Science* 262, no 5, (May 2003): 76 - 81.

United States Air Force, Air University. *Spacecast 2020 Executive Summary*. 23 February 1998. <http://www.au.af.mil/Spacecast/monographs/exec-sum.pdf> (2004).

University of Tennessee. "Newtonian Gravitation and the Laws of Kepler," 4 October 2004. <http://csep10.phys.utk.edu/astr161/lect/history/newtonkepler.html> (2004).

Walker, Roy, Major, USAF, and Larry Ridolfi, Captain, USAF. "Airpower's Role in Maritime Operations." *Air & Space Power Chronicles.* <http://www.airpower.maxwell.af.mil/airchronicles/cc/ridolfi.html> (2004).

Walling, Eileen M. "High-Power Microwaves and Modern Warfare." In *The Technological Arsenal*, ed. William C. Martel, 90-106. Washington: Smithsonian Institution Press, 2001.

Worden, Simon P. "Space Control in the 21st Century: A Space 'Navy' Protecting the Commercial Basis of America's Wealth." In *Spacepower for a New Millennium*, ed. Peter L. Hays, James M. Smith, Alan R. Van Tassel, and Guy M. Walsh, 225-238. New York: McGraw-Hill, 2000.

X Prize Foundation. "What is the X PRIZE™?" <http://www.xprize.org/press/what.html> (2004).

About the Author

Wilson Wong is Research Associate for the Aerospace Research Policy Group, Winnipeg, Manitoba, and Research Assistant for the Centre for Defence and Security Studies. He has Master or Arts degree in Political Studies from the University of Manitoba.